Undergraduate Texts in Mathematics

Undergraduate Texts in Mathematics

Undergraduate Texts in Mathematics are generally aimed at third- and fourth-year undergraduate mathematics students at North American universities. These texts strive to provide students and teachers with new perspectives and novel approaches. The books include motivation that guides the reader to an appreciation of interrelations among different aspects of the subject. They feature examples that illustrate key concepts as well as exercises that strengthen understanding.

More information about this series at http://www.springer.com/series/666

Daniel Rosenthal • David Rosenthal
Peter Rosenthal

A Readable Introduction
to Real Mathematics

Second Edition

Daniel Rosenthal
Toronto, ON, Canada

Peter Rosenthal
Department of Mathematics
University of Toronto
Toronto, ON, Canada

David Rosenthal
Department of Mathematics
and Computer Science
St. John's University
Queens, NY, USA

ISSN 0172-6056   ISSN 2197-5604   (electronic)
Undergraduate Texts in Mathematics
ISBN 978-3-030-80731-3   ISBN 978-3-030-00632-7   (eBook)
https://doi.org/10.1007/978-3-030-00632-7

Library of Congress Control Number: 2018959426

Mathematics Subject Classification (2010): 03E10, 11A05, 11A07, 11A41, 11A51, 11-01, 51-01, 97-01, 97F30, 97F40, 97F50, 97F60, 97G99, 97H99

This Springer imprint is published by the registered company Springer Nature Switzerland AG.
The registered company address is: Gewerbestrasse 11, 6330 Cham, Switzerland

To the memory of Harold and Esther Rosenthal who gave us (and others) the gift of mathematics.

Preface to the Second Edition

This second edition is an expanded and improved version of the first. The last two chapters are entirely new. The other chapters have been revised, taking into account the comments of many readers. We are particularly grateful to Florin Catrina, Jonathan Korman, Andrew Nicas, Carolyn Pitchik, Heydar Radjavi and Zack Wolske for their suggestions.

The preface to the first edition has been rewritten and divided into two prefaces, one for readers and one for instructors.

There are undoubtedly further improvements that could be made. We would appreciate your sending any comments, corrections, or suggestions to any of the authors at their e-mail addresses given below.

Daniel Rosenthal: danielkitairosenthal@gmail.com
David Rosenthal: rosenthd@stjohns.edu
Peter Rosenthal: rosent@math.toronto.edu

Toronto, ON, Canada   Daniel Rosenthal
Queens, NY, USA   David Rosenthal
Toronto, ON, Canada   Peter Rosenthal

Preface for Readers

The fundamental purpose of this book is to teach you to understand mathematical thinking. We have tried to do that in a way that is clear and engaging, and emphasizes the beauty of mathematics. You may be reading this book on your own or as a text for a course you are enrolled in. Regardless of your reason for reading this book, we hope that you will find it understandable and interesting.

This book contains a lot of mathematics. We do not expect you to necessarily read all of it. In the Preface for Instructors, we describe possible courses that use only parts of the book.

Mathematics is a huge and growing body of knowledge; no one can learn more than a fraction of it. But the central thing to learn is how to think mathematically. It is our experience that mathematical thinking can be learned by almost anyone who is willing to make a serious attempt. We invite you to make such an attempt by reading at least part of this book. It is important not to let yourself be discouraged if you can't easily understand something. Everyone learning mathematics finds some concepts baffling at first, but usually, with enough effort, the ideas become clear.

One way in which mathematics gets very complex is by building on itself; some mathematical concepts are built on a foundation of many other concepts and thus require a great deal of background to understand. That is not the case for the topics discussed in this book. Reading this book does not require any background other than basic high school algebra and, for parts of Chapters 9 and 12, some high school trigonometry.

A few questions, among the many, that you will easily be able to answer after reading the relevant parts of this book are the following: Is $13^{217} \cdot 37^{92} \cdot 41^{15} = 19^{111} \cdot 29^{145} \cdot 43^{12} \cdot 47^5$ (see Chapter 4)? Is there a largest prime number (i.e., a largest whole number whose only factors are 1 and itself) (Theorem 1.1.5)? If a store sells one kind of product for 9 dollars each and another kind for 16 dollars each and receives 143 dollars for the total sale of both, how many products did the store sell at each price (Example 7.2.7)? How do computers send secret messages to each other (Chapter 6)? How is the size of an infinite set defined? Are there more fractions than there are whole numbers? Are there more real numbers than there are fractions? Is there a smallest infinity? Is there a largest infinity (Chapter 10)? What

Preface for Readers

are complex numbers (Chapter 9)? Is .3333 ... really equal to $\frac{1}{3}$ (Example 13.2.8)? What are some infinite-dimensional spaces (Section 14.5)?

The hardest theorem proven in this book concerns the construction of angles using a compass and a straightedge. (A straightedge is a ruler-like device but without measurements marked on it. Straightedges are used to draw lines connecting two points.) If you are given any angle, it is easy to bisect it (i.e., divide it into two equal subangles) by using a compass and a straightedge (we will show you how to do that). This and many similar results were discovered by the Ancient Greeks. The Ancient Greeks wondered whether angles could be "trisected" in the sense of being divided into three equal subangles using only a straightedge and a compass. A lot of mathematics beyond that conceived of by the Ancient Greeks was required to solve this problem; it was not solved until the nineteenth century. It can be proven that many angles, including an angle of 60 degrees, cannot be so trisected. We present a complete proof of this as an illustration of complicated but beautiful mathematical reasoning.

The most important question you'll be able to answer after reading at least several chapters of this book, although you will have difficulty formulating the answer in words, is: what is mathematical thinking really like? If you read and understand several chapters and do a fair number of the problems that are provided, you will certainly have a feeling for mathematical thinking.

We hope that you read this book carefully. Reading mathematics is not like reading a novel, a newspaper, or anything else. As you go along, you have to really reflect on the mathematical reasoning that is being presented. After reading a description of an idea, think about it. When reading mathematics you should always have a pencil and paper at hand to rework what you read.

The essence of mathematics consists of theorems, which are statements proven to be true. We will prove a number of theorems. When you begin reading about a theorem, think about why it may be true before you read our proof. In fact, at some points you may be able to prove the theorem we state without looking at our proof at all. In any event, you should make at least a small attempt before reading the proof in the book. It is often useful to continue such attempts while in the middle of reading the proof that we present; once we have gotten you a certain way towards the result, see if you can continue on your own.

If you adopt such an approach and are patient, we believe that you will learn to think mathematically. We are also convinced that you will feel that much of the mathematics that you learn is beautiful, in the sense that you will find that the logical argument that establishes the theorem is what mathematicians call "elegant."

We chose the material for this book based on the following criteria: the mathematics is beautiful, it is useful in many mathematical contexts, and it is accessible without much mathematical background. The theorems that we prove have applications to mathematics and to problems in other subjects.

Each chapter ends with a section entitled "Problems." The "Problems" sections are divided into three subsections. You should do some of the "Basic Exercises" to ensure that you have an understanding of the fundamentals of the chapter.

The subsections entitled "Interesting Problems" contain problems whose solutions depend upon the material of the chapter and seem to have mathematical or other interest. The subsections labeled "Challenging Problems" contain problems that we expect you will, indeed, find to be quite challenging. You should not be discouraged if you cannot solve some of the problems. However, if you do solve problems that you find difficult at first, especially those that we have labeled "challenging," we hope and expect that you will experience some of the pleasure and satisfaction that mathematicians feel upon discovering new mathematics.

Each chapter is divided into sections. Important items, such as definitions and theorems, are numbered in a way that locates them within a chapter and a section of that chapter. We put the chapter number, then the section number, and then the number of the item within that section. For example, 7.2.4 refers to the fourth numbered item in section two of chapter seven.

Readers who wish to omit some of the material (perhaps only at first) should be aware of the following. Chapters 1, 2, 4, and 8 may be read without reading any other parts of this book. Chapter 5 depends on Chapters 3 and 4, and Chapter 6 requires Chapter 5. Some of the examples in Chapter 7 depend on Chapter 6; the rest of the chapter is independent of Chapter 6. Chapter 8 uses Chapter 4. Chapters 9, 10, and 11 are essentially independent of each other and of all other chapters. Chapter 12 basically depends only on Chapter 11 and on the concepts of rational and irrational numbers as discussed in Chapter 8. Chapter 13 can be read independently of the other chapters, and Chapter 14 does not require any of the previous material except for the essential properties of convergence of infinite series, as discussed in Chapter 13.

Preface for Instructors

A glance at the table of contents and the Preface for Readers will give you an idea of the material covered in this text.

Some features of this book include:

- Complete proofs that an angle of 60 degrees cannot be trisected with a straightedge and compass (Corollary 12.3.24) and that an angle of an integral number n degrees can be constructed with a straightedge and compass if and only if n is a multiple of 3 (Theorem 12.4.13).
- A thorough discussion of the Principle of Mathematical Induction (Chapter 2).
- A chapter that provides an introduction to Euclidean plane geometry (Chapter 11).
- A complete description of RSA encryption (7.2.5).
- A fairly extensive treatment of cardinality (Chapter 10).
- An introduction to infinite-dimensional spaces (Chapter 14).
- Using the least upper bound property to establish theorems about convergence of infinite series (Section 13.4).
- Showing that real numbers can be represented by infinite decimals (Section 13.6).
- A proof that the infinite series consisting of the reciprocals of the prime numbers diverges (Theorem 13.7.8).

Since the only prerequisite for understanding this book is high school algebra, it is suitable as a textbook for a wide variety of courses. In particular, it is our view that appropriate parts of the text could be used for courses for mathematics or science majors, for courses for other students who want to get an appreciation of mathematics, and for courses for prospective teachers. The book is also written so as to be useable for independent study by anyone who is interested in learning mathematics. In particular, mathematically inclined high school students might be directed to this book.

The main purpose of this book is to teach mathematical thinking. Some instructors like to begin such a course by discussing basic logic and different kinds of proofs. Others prefer to present some interesting mathematics simply and clearly, with the expectation that students will learn to think mathematically by being gently exposed to the mathematics presented.

We are in the latter camp.

The text begins with a basic introduction to the natural numbers. This is followed by a chapter that contains a thorough discussion of mathematical induction. A student who has learned to understand most of the material in that chapter will have obtained some appreciation of mathematical thinking. Learning the material in other parts of the book will deepen the student's understanding and will also teach the student a lot of interesting mathematics. The textbook provides the opportunity for you to choose from a variety of mathematical topics.

The following are some descriptions of different courses for which part or all of this book could serve as a text. There are many other variants that instructors could devise.

A course covering most of the book would take two semesters. Such a course would be suitable for students majoring in mathematics, statistics, computer science, or physics.

On the other hand, there are several different one-semester courses that could be based on parts of the book. Instructors can vary the level of these courses by the pace at which they proceed, the difficulty of the problems that they assign, and the material they omit.

One natural possibility would be to begin at page 1, proceed at whatever pace is comfortable for you and your students, and then see where you end up.

Other possibilities involve omitting some of the chapters. It is our opinion that Chapters 1, 2, 4, and 8 should be part of most courses using this book. Additional chapters can be chosen based on the needs of the students, the interests of the instructor, and the time available. For example, Chapters 3, 5, 6, and 7 could be added (i.e., so that the course covers Chapters 1 through 8). Alternatively, Chapters 10 and 13 and/or 14 might be included.

A course containing a proof that some angles cannot be trisected with straight-edge and compass could be based on Chapters 1, 2, 4, 8, 11, and 12. Other chapters could be added if time permits.

Fairly leisurely "mathematics appreciation" courses could cover Chapters 1, 2, 3, 4, and 5, or Chapters 1, 2, 4, 8, and 10, or Chapters 1, 2, 4, 8, and 13, or Chapters 1, 2, 4, 8, and the part of Chapter 14 before Definition 14.5.3.

A one-semester course for prospective or actual teachers of high school mathematics could cover Chapters 1, 2, 4, 8, 11, and 12. It is our view that Chapter 12 should be of substantial interest to teachers. If they are already familiar with the fundamentals of Euclidean geometry as presented in Chapter 11, there would likely be time to add one or more of Chapters 9, 10, 13, or 14. If the instructor does not wish to present Chapter 12, a good course for teachers could be based on Chapters 1 through 8.

Lectures on parts of some of the chapters can give students a taste of the topic. For example, a very brief introduction to cardinality could consist of the part of Chapter 10 up through Theorem 10.2.3. Chapter 14 describes some finite and infinite-dimensional spaces. The part before Definition 14.5.3 is completely independent of the rest of the book; the balance requires the concept of convergence of series.

Chapter 11 is an essentially self-contained introduction to geometry.

Chapter 13, which does not significantly rely on any other chapters, is intended to provide a first introduction to concepts of analysis by explaining convergence of infinite series in a direct manner. The idea of "adding lots of terms to get close to the sum" has some intuitive appeal. In our experience, many students who are taught the traditional approach are confused by the distinction between convergence of the sequence of terms and convergence of the sequence of partial sums. That is why we do not discuss convergence of any sequences other than sequences of partial sums. Also, we do not use "sigma notation" within the chapter since some students find it to be a barrier to understanding. We define least upper bounds and use that concept to rigorously prove the fundamental theorems about convergence. The connections to the more standard approaches are established in the last problem of the chapter.

Using a text that contains more than will be covered in the course you are teaching provides an opportunity to encourage interested students to do some reading on their own, before or after the course ends.

Contents

Chapter 1
Introduction to the Natural Numbers

We assume basic knowledge about the numbers that we count with; that is, the numbers 1, 2, 3, 4, 5, 6, and so on. These are called the *natural numbers*, and the collection of all of them is usually denoted by $\mathbb{N}$. They do seem to be very natural, in the sense that they arose very early on in virtually all societies. There are many other names for these numbers, such as the *positive integers* and the *positive whole numbers*. Although the natural numbers are very familiar, we will see that they have many interesting properties beyond the obvious ones. Moreover, there are many questions about the natural numbers to which nobody knows the answer. Some of these questions can be stated very simply, as we shall see, although their solutions have eluded the thousands of mathematicians who have attempted to solve them.

We assume familiarity with the two basic operations on the natural numbers, addition and multiplication. The sum of two numbers will be indicated using the plus sign "+". Multiplication will be indicated by putting a dot in the middle of the line between the numbers, or by simply writing the symbols for the numbers next to each other, or sometimes by enclosing them in parentheses. For example, the product of 3 and 2 could be denoted $3 \cdot 2$ or $(3)(2)$. The product of the natural numbers represented by the symbols m and n could be denoted mn, or $m \cdot n$, or $(m)(n)$.

We also, of course, need the number 0. Moreover, we require the negative whole numbers as well. For each natural number n there is a corresponding negative number $-n$ such that $n + (-n) = 0$. Altogether, the collection of positive and negative whole numbers and 0 is called the *integers*. It is often denoted by $\mathbb{Z}$.

We assume that you know how to add two negative integers and also how to add a negative integer to a positive integer. Multiplication appears to be a bit more mysterious. Most people feel comfortable with the fact that, for m and n natural numbers, the product of m and $(-n)$ is $-mn$. What some people find more mysterious is the fact that $(-m)(-n) = mn$ for natural numbers m and n; that is, the product of two negative integers is a positive integer. There are various possible explanations that can be provided for this, one of which is the following. Using the

© Springer Nature Switzerland AG 2018
D. Rosenthal et al., *A Readable Introduction to Real Mathematics*,
Undergraduate Texts in Mathematics, https://doi.org/10.1007/978-3-030-00632-7_1

usual rules of arithmetic:

$$0 = (-m)(0) = (-m)(-n + n) = (-m)(-n) + (-m)(n)$$

Adding mn to both sides of this equation gives

$$0 + mn = (-m)(-n) + (-m)(n) + mn$$

or

$$mn = (-m)(-n) + \big((-m) + m\big) \cdot n$$

Thus,

$$mn = (-m)(-n) + 0 \cdot n$$

so

$$mn = (-m)(-n)$$

Therefore, the fact that $(-m)(-n) = mn$ is implied by the other standard rules of arithmetic.

1.1   Prime Numbers

One of the important concepts we will study is *divisibility*. For example, 12 is divisible by 3, which means that there is a natural number (in this case, 4) such that the product of 3 and that natural number is 12. That is, $12 = 3 \cdot 4$. In general, we say:

Definition 1.1.1.   The integer m is *divisible* by the integer n if there exists an integer q such that $m = nq$.

There are many other terms that are used to describe such a relationship. For example, if $m = nq$, we may say that n and q are *divisors* of m and that each of n and q *divides* m. Note that every integer divides 0, since $0 = n \cdot 0$ for every integer n. The terminology "q is the quotient when m is divided by n" is also used when n is different from 0. In this situation, n and q are also sometimes called *factors* of m; the process of writing an integer as a product of two or more integers is called *factoring* the integer.

The number 1 is a divisor of every natural number since, for each natural number m, $m = 1 \cdot m$. Also, every natural number m is a divisor of itself, since $m = m \cdot 1$.

The number 1 is the only natural number that has only one natural number divisor, namely itself. Every other natural number has at least two divisors, itself and 1.

Definition 1.1.2. A *prime number* is a natural number greater than 1 whose only natural number divisors are 1 and the number itself.

The first prime number is 2. The primes continue: 3, 5, 7, 11, 13, 17, 19, 23, 29, 31, and so on.

And so on? Is there a largest prime? Or does the sequence of primes continue without end? There is, of course, no largest natural number. For if n is any natural number, then $n + 1$ is a natural number and $n + 1$ is bigger than n. It is not so easy to determine if there is a largest prime number or not. If p is a prime, then $p + 1$ is almost never a prime. If $p = 2$, then $p + 1 = 3$ and p and $p + 1$ are both primes. However, 2 is the only prime number p for which $p + 1$ is prime. This can be proven as follows. First note that, since every even number is divisible by 2, 2 itself is the only even prime number. Therefore, if p is a prime other than 2, then p is odd and $p + 1$ is an even number larger than 2 and is thus not prime.

Is it nonetheless true that, given any prime number p, there is a prime number larger than p? Although we cannot get a larger prime by simply adding 1 to a given prime, there may be some other way of establishing that there is a prime number larger than any given one. We will answer this question after learning a little more about primes.

A natural number, other than 1, that is not prime is said to be *composite*. (The number 1 is special and is neither prime nor composite.) For example, 4, 68, 129, and 2010 are composites. Thus, a composite number is a natural number that has a divisor in addition to itself and 1.

To determine if a number is prime, what potential factors must be checked to eliminate the possibility that there are factors other than the number and 1? Fortunately, to check whether or not a natural number m is prime, you need not check whether every natural number less than m divides m.

Theorem 1.1.3. *Let m be a natural number other than 1. If m does not have a natural number divisor that is greater than 1 and no larger than the square root of m, then it is prime.*

Proof. If $m = n \cdot q$, it is not possible that n and q are both larger than the square root of m, for if two natural numbers are both larger than the square root of m, then their product is larger than m. It follows that a natural number greater than 1 that is not prime has at least one divisor that is larger than 1 and is no larger than the square root of that natural number. □

For example, we can conclude that 101 is prime since none of the numbers 2, 3, 4, 5, 6, 7, 8, 9, 10 are divisors of 101.

Using sophisticated techniques and computers, many very large numbers have been shown to be prime. For example, 100,000,561 is prime, as is 22,801,763,489.

The fact that very large natural numbers have been shown to be prime does not answer the question of whether there is a largest prime. The theorem that there is always a prime larger than p for *every* prime number p cannot be established by computing any number of specific primes, no matter how large.

Over the centuries, mathematicians have discovered many proofs that there is no largest prime. We shall present one of the simplest and most beautiful proofs, discovered by the Ancient Greeks.

We begin by establishing a preliminary fact that is required for the proof. A statement that is proven for the purpose of being used to prove something else is called a "lemma." We need a lemma. The lemma that we require states that every composite number has a divisor that is a prime number. (The proof that we present of the lemma is quite convincing, but we shall subsequently present a more precise proof; see Lemma 2.2.3.)

Lemma 1.1.4. *Every natural number greater than 1 has a prime divisor.*

Proof. If the given natural number is prime, then it is a prime divisor of itself. If the number, say m, is composite, then m has at least one factorization $m = n \cdot q$ where neither n nor q is m or 1. If either of n or q is a prime number, then the lemma is established for m. If n is not prime, then it has a factorization $n = s \cdot t$, where s and t are natural numbers other than 1 and n. It is clear that s and t are also divisors of m. Thus, if either of s and t is a prime number, the lemma is established. If s is not prime, then it can be factored into a product where neither factor is s or 1, and so on. Continued factoring must get down to a factor that cannot itself be factored; i.e., to a factor that is prime. That prime number is a divisor of m, so the lemma is established.   □

The following is the ingenious proof of the infinitude of the primes discovered by the Ancient Greeks.

Theorem 1.1.5. *There is no largest prime number.*

Proof. Let p be any prime number. We must prove that there is some prime larger than p. To do this, we will construct a number that we will show is either a prime larger than p or has a prime divisor larger than p. In both cases, we will conclude that there is a prime number larger than p.

Here is how we construct the large number. Let M be the number obtained by taking the product of all the prime numbers up to and including the given prime p and then adding 1 to that product. That is,

$$M = (2 \cdot 3 \cdot 5 \cdot 7 \cdot 11 \cdot 13 \cdot 17 \cdot 19 \cdots p) + 1$$

It is possible that M is a prime number. If that is so, then there is a prime number larger than p, since M is obviously larger than p. If M is not prime, then it is composite. We must show that there is a prime larger than p in this case as well.

Suppose, then, that M is composite. By Lemma 1.1.4, it follows that M has a prime divisor. Let q be any prime divisor of M. We will show that q is larger than p and thus that there is a prime larger than p in this case as well.

Consider possible values of q, a prime divisor of M. Surely q is not 2, for

$$2 \cdot 3 \cdot 5 \cdot 7 \cdot 11 \cdot 13 \cdot 17 \cdot 19 \cdots p$$

is an even number, and thus adding 1 to that number to get M produces an odd number. That is, M is odd and is therefore not divisible by 2. Since q does divide M, q cannot be equal to 2.

Similar reasoning shows that q cannot be 3. For

$$2 \cdot 3 \cdot 5 \cdot 7 \cdot 11 \cdot 13 \cdot 17 \cdot 19 \cdots p$$

is a multiple of 3, so the number obtained by adding 1, namely M, leaves a remainder of 1 when it is divided by 3. That is, 3 is not a divisor of M. Since q is a divisor of M, q is not 3.

Exactly the same proof shows that q is not 5, since 5 is a divisor of

$$2 \cdot 3 \cdot 5 \cdot 7 \cdot 11 \cdot 13 \cdot 17 \cdot 19 \cdots p$$

and thus cannot be a divisor of M. In fact, the same proof establishes that q cannot be any of the factors $2, 3, 5, \ldots, p$ of the product

$$2 \cdot 3 \cdot 5 \cdot 7 \cdot 11 \cdot 13 \cdot 17 \cdot 19 \cdots p$$

Since every prime number up to and including p is a factor of that product, q cannot be any of those prime numbers. Therefore q is a prime number that is not any of the prime numbers up to and including p. It follows that q is a prime number larger than p, and we have proven that there is a prime number larger than p in the case where M is composite. Therefore, in both cases, the case where M is prime and the case where M is composite, we have shown that there is a prime number larger than p. This proves the theorem. $\qquad\square$

Every mathematician would agree that the above proof is "elegant." If you find the proof interesting, then you are likely to appreciate many of the other ideas that we will discuss (and much mathematics that we do not cover as well).

1.2 Unanswered Questions

There are many questions concerning prime numbers that no one has been able to answer. One famous question concerns what are called *twin primes*. Since 2 is the only even prime number, the only consecutive integers that are prime are 2 and 3. There are, however, many pairs of primes that are two apart, such as

{3, 5}, {29, 31}, {101, 103}, {1931, 1933}, and {104471, 104473}. Such pairs are called *twin primes*. One question that remains unanswered, in spite of the efforts of thousands of mathematicians over hundreds of years, is the question of whether there is a largest pair of twin primes. Some very large pairs are known (e.g., {1000000007, 1000000009} and many pairs that are even much bigger), but no one knows if there is a largest such.

Another very famous unsolved problem is whether or not the *Goldbach Conjecture* is true. Several hundred years ago, Goldbach conjectured (that is, said that he thought that it was probably true) that every even natural number larger than 2 is the sum of two prime numbers (e.g., $6 = 3 + 3$, $20 = 7 + 13$, and $22,901,764,050 = 22,801,763,489 + 100,000,561$). Goldbach's Conjecture is known to be true for many very large even natural numbers, but no one has been able to prove it in general (or to show that there is an even number that cannot be written as the sum of two primes).

If you are able to solve the twin primes problem or determine the truth or falsity of Goldbach's Conjecture, you will immediately become famous throughout the world and your name will remain famous as long as civilization endures. On the other hand, it will almost undoubtedly prove to be extremely difficult to answer either of those questions. On the other "other hand," there is a very slight possibility that one or both of those questions have a fairly simple answer that has been overlooked by the many great and not-so-great mathematicians who have thought about them. In spite of the very small possibility of success, you might find it interesting to think about these problems.

1.3   Problems

Basic Exercises

1. Show that the following are composite numbers:

   (a) 68
   (b) 129
   (c) 20,101,116

2. Which of the following are prime numbers?

   (a) 79
   (b) 153
   (c) 537
   (d) 851,486

3. Write each of the following numbers as a sum of two primes:

   (a) 100
   (b) 112

Interesting Problems

4. Verify that the Goldbach Conjecture holds for all even numbers between 4 and 50.
5. Find a pair of twin primes such that each prime is greater than 200.

Challenging Problems

6. Find a prime number p such that the number $(2 \cdot 3 \cdot 5 \cdot 7 \cdots p) + 1$ is not prime.
7. Suppose that p, $p + 2$, and $p + 4$ are prime numbers. Prove that $p = 3$.
   [Hint: Why can't p be 5 or 7?]
8. Prove that, for every natural number n greater than 2, there is a prime number between n and $n!$. (Recall that, for every natural number n, $n!$, which is read "n factorial", denotes the product of all the natural numbers from 1 up to n.)
   [Hint: There is a prime number that divides $n! - 1$.]
   Note that this gives an alternate proof that there are infinitely many prime numbers.
9. Prove that, for every natural number n, there are n consecutive composite numbers.
   [Hint: $(n + 1)! + 2$ is a composite number.]
10. Show that a natural number has an odd number of different factors if and only if it is a perfect square (i.e., it is the square of another natural number).

Chapter 2
Mathematical Induction

There is a method for proving certain theorems that is called *mathematical induction*. We will give a number of examples of proofs that use this method. The basis for mathematical induction, however, is a statement about sets of natural numbers. We use the word *set* informally to mean any collection of things, and each "thing" is said to be an *element* of the set. (For more on sets, see Chapter 10.) Recall that the set of all natural numbers is the set $\{1, 2, 3, \dots\}$. Mathematical induction provides an alternate description of that set.

2.1 The Principle of Mathematical Induction

The way mathematical induction is usually explained can be illustrated by considering the following example. Suppose that we wish to prove, for every natural number n, the validity of the following formula for the sum of the first n natural numbers:

$$1 + 2 + 3 + \cdots + (n - 1) + n = \frac{n(n + 1)}{2}$$

One way to prove that this formula holds for every n is the following. First, the formula does hold for $n = 1$, for in this case the left-hand side is just 1 and the right-hand side is $\frac{1 \cdot (1+1)}{2}$, which is equal to 1. To prove that the formula holds for all n, we will establish the fact that whenever the formula holds for any given natural number, the formula will also hold for the next natural number. That is, we will prove that the formula holds for $n = k + 1$ whenever it holds for $n = k$. (This passage from k to $k + 1$ is often called "the inductive step.") If we prove this fact, then, since we know that the formula does hold for $n = 1$, it would follow from this fact that it holds for the next natural number, 2. Then, since it holds for $n = 2$, it

© Springer Nature Switzerland AG 2018
D. Rosenthal et al., *A Readable Introduction to Real Mathematics*,
Undergraduate Texts in Mathematics, https://doi.org/10.1007/978-3-030-00632-7_2

holds for the natural number that follows 2, which is 3. Since it holds for 3, it holds
for 4, and then for 5, and 6, and so on. Thus, we will conclude that the formula holds
for every natural number.

To prove the formula in general, then, we must show that the formula holds for
$n = k + 1$ whenever it holds for $n = k$. Assume that the formula does hold for
$n = k$, where k is any fixed natural number. That is, we assume the formula

$$1 + 2 + 3 + \cdots + (k - 1) + k = \frac{k(k + 1)}{2}$$

We want to derive the formula for $n = k + 1$ from the above equation. We do that
as follows. Assuming the above formula, add $k + 1$ to both sides, getting

$$1 + 2 + 3 + \cdots + (k - 1) + k + (k + 1) = \frac{k(k + 1)}{2} + (k + 1)$$

We shall see that a little algebraic manipulation of the right-hand side of the above
will produce the formula for $n = k + 1$. To see this, note that

$$\frac{k(k + 1)}{2} + (k + 1) = \frac{k(k + 1)}{2} + \frac{2(k + 1)}{2}$$
$$= \frac{k(k + 1) + 2(k + 1)}{2}$$
$$= \frac{(k + 2)(k + 1)}{2}$$
$$= \frac{(k + 1)(k + 2)}{2}$$
$$= \frac{(k + 1)\big((k + 1) + 1\big)}{2}$$

Thus, $1 + 2 + 3 + \cdots + (k - 1) + k + (k + 1) = \frac{(k+1)((k+1)+1)}{2}$. This equation is
the same as that obtained from the formula by substituting $k + 1$ for n. Therefore
we have established the inductive step, so we conclude that the formula does hold
for all n.

The Principle of Mathematical Induction, which is implicitly used in the above
proof, is really just an assertion about sets of natural numbers.

Suppose S is a set of natural numbers that has the following two properties:

A. *The number 1 is in S.*
B. *Whenever a natural number is in S, the next natural number is also in S.*

The second property can be stated a little more formally: If k is a natural number
and k is in S, then $k + 1$ is in S.

What can we say about a set S that has those two properties? Since 1 is in S (by
property A), it follows from property B that 2 is in S. Since 2 is in S, it follows from
property B that 3 is in S. Since 3 is in S, 4 is in S. Then 5 is in S, 6 is in S, 7 is in
S, and so on. It seems clear that S must contain every natural number. That is, the
only set of natural numbers with the above two properties is the set of all natural
numbers. We state this formally:

The Principle of Mathematical Induction 2.1.1. *If S is any set of natural numbers with the properties that*

A. *1 is in S, and*
B. *k + 1 is in S whenever k is any number in S,*

then S is the set of all natural numbers.

In introducing the Principle of Mathematical Induction, we gave an indication of why it is true. A more formal proof can be based on the following more obvious fact, which we assume as an axiom.

The Well-Ordering Principle 2.1.2. *Every set of natural numbers that contains at least one element has a smallest element in it.*

We can establish the Principle of Mathematical Induction from the Well-Ordering Principle as follows. Suppose that the Well-Ordering Principle holds for all sets of natural numbers. Let S be any set of natural numbers and suppose that S has properties A and B of the Principle of Mathematical Induction. To prove the Principle of Mathematical Induction, we must prove that the only such set S is the set of all natural numbers. We will do this by showing that it is impossible that there is any natural number that is not in S. To see this, suppose that S does not contain all natural numbers. Then let T denote the set of all natural numbers that are not in S. Assuming that S is not the set of all natural numbers is equivalent to assuming that T has at least one element. If this were the case, then well-ordering would imply that T has a smallest element. We will show that this is impossible.

Suppose that t was the smallest element of T. Since 1 is in S, 1 is not in T. Therefore, t is larger than 1. It follows that $t - 1$ is a natural number. Since $t - 1$ is less than the smallest number t in T, $t - 1$ cannot be in T. Since T contains all the natural numbers that are not in S, this implies that $t - 1$ is in S. This, however, leads to the following contradiction. Since S has property B, $(t - 1) + 1$ must also be in S. But this is t, which is in T and therefore not in S. Therefore the assumption that there is a smallest element of T is not consistent with the properties of S. Thus, there is no smallest element of T and, by well-ordering, there is therefore no element in T. This proves that S is the set of all natural numbers.

The way the formal principle applies to examples such as the one given above is by letting the set S be the set of natural numbers for which the formula holds. Showing that S is the set of all natural numbers is equivalent to showing that the formula holds for every natural number.

There are many very similar proofs of similar formulas.

Theorem 2.1.3. *For every natural number n,*

$$1^2 + 2^2 + 3^2 + \cdots + n^2 = \frac{n(n+1)(2n+1)}{6}$$

Proof. Let S be the set of all natural numbers for which the formula is true. We want to show that S contains all of the natural numbers. We do this by showing that S has properties A and B.

For property A, we need to check that $1^2 = \frac{1(1+1)(2\cdot1+1)}{6}$. This is true, so S satisfies property A. To verify property B, let k be in S. We must show that $k + 1$ is in S. Since k is in S, the formula holds for k. That is,

$$1^2 + 2^2 + 3^2 + \cdots + k^2 = \frac{k(k+1)(2k+1)}{6}$$

Using this formula, we can prove the corresponding formula for $k + 1$ as follows. Adding $(k + 1)^2$ to both sides of the above equation, we get

$$1^2 + 2^2 + 3^2 + \cdots + k^2 + (k+1)^2 = \frac{k(k+1)(2k+1)}{6} + (k+1)^2$$

Now we do some algebraic manipulations to the right-hand side to see that it is what we want:

$$\frac{k(k+1)(2k+1)}{6} + (k+1)^2 = \frac{k(k+1)(2k+1) + 6(k+1)^2}{6}$$

$$= \frac{(k+1)\Big(k(2k+1) + 6(k+1)\Big)}{6}$$

$$= \frac{(k+1)\Big((2k^2+k) + (6k+6)\Big)}{6}$$

$$= \frac{(k+1)(2k^2 + 7k + 6)}{6}$$

$$= \frac{(k+1)(k+2)(2k+3)}{6}$$

The last equation is the formula in the case when $n = k + 1$, so $k + 1$ is in S. Therefore, S is the set of all natural numbers by the Principle of Mathematical Induction.   □

Sometimes one wants to prove something by induction that is not true for all natural numbers, but only for those bigger than a given natural number. A slightly more general principle that can be used in such situations is the following.

The Generalized Principle of Mathematical Induction 2.1.4. *Let m be a natural number. If S is a set of natural numbers with the properties that*

A. *m is in S, and*
B. *$k + 1$ is in S whenever k is in S and is greater than or equal to m,*

then S contains every natural number greater than or equal to m.

   The Principle of Mathematical Induction is the special case of the generalized principle when $m = 1$. The generalized principle states that we can use induction starting at any natural number, not just at 1.

   For example, consider the question: which is larger, $n!$ or 2^n? (Recall that $n!$ is the product of the natural numbers from 1 up to n.) For $n = 1, 2$, and 3, we see that

$$1! = 1 < 2^1 = 2$$

$$2! = 2 \cdot 1 = 2 < 2^2 = 2 \cdot 2 = 4$$

$$3! = 3 \cdot 2 \cdot 1 = 6 < 2^3 = 2 \cdot 2 \cdot 2 = 8$$

But when $n = 4$, the inequality is reversed, since

$$4! = 4 \cdot 3 \cdot 2 \cdot 1 = 24 > 2^4 = 2 \cdot 2 \cdot 2 \cdot 2 = 16$$

When $n = 5$,

$$5! = 5 \cdot 4 \cdot 3 \cdot 2 \cdot 1 = 120 > 2^5 = 2 \cdot 2 \cdot 2 \cdot 2 \cdot 2 = 32$$

If you think about it a bit, it is clear why eventually $n!$ is much bigger than 2^n. In both expressions we are multiplying n numbers together, but for 2^n we are always multiplying by 2, whereas the numbers we multiply to build $n!$ get larger and larger. While it is not true that $n! > 2^n$ for every natural number (since it is not true when n is 1, 2, or 3), we can, as we now show, use the more general form of mathematical induction to prove that it is true for all natural numbers greater than or equal to 4.

Theorem 2.1.5. $n! > 2^n$ *for* $n \geq 4$.

Proof. We use the Generalized Principle of Mathematical Induction with $m = 4$. Let S be the set of natural numbers for which the theorem is true. As we saw above, $4! > 2^4$. Therefore, 4 is in S. Thus, property A is satisfied. For property B, assume that $k \geq 4$ and that k is in S; i.e., $k! > 2^k$. We must show that $(k + 1)! > 2^{k+1}$. Multiplying both sides of the inequality for k (which we have assumed to be true) by $k + 1$ gives

$$(k + 1)(k!) > (k + 1) \cdot 2^k$$

The left-hand side is just $(k + 1)!$; therefore we have the inequality

$$(k + 1)! > (k + 1) \cdot 2^k$$

Since $k \geq 4$, $k + 1 > 2$. Therefore, the right-hand side of the inequality, $(k + 1) \cdot 2^k$, is greater than $2 \cdot 2^k = 2^{k+1}$. Combining this with the above inequality, we get

$$(k + 1)! > (k + 1) \cdot 2^k > 2^{k+1}$$

Thus, $k + 1$ is in S, which verifies property B. By the Generalized Principle of Mathematical Induction, S contains all natural numbers $n \geq 4$.   □

The following is an example where mathematical induction is useful in establishing a geometric result. We will use the word "tromino" to denote an L-shaped object consisting of three squares of the same size. That is, a tromino looks like this:

Another way to think of a tromino is that it is the geometric figure obtained by taking a square that is composed of four smaller squares and removing one of the smaller squares.

We are going to consider what geometric regions can be covered by trominos, all of which have the same size and do not overlap each other. As a first example, start with a square made up of 16 smaller squares (i.e., a square that is "4 by 4") and remove one small square from a corner of the square:

Can the region that is left be covered by trominos (each made up of three small squares of the same size as the small squares in the region) that do not overlap each other? It can:

We can use mathematical induction to prove the following.

Theorem 2.1.6. *For each natural number n, consider a square consisting of 2^{2n} smaller squares. (That is, a $2^n \times 2^n$ square.) If one of the smaller squares is removed from a corner of the large square, then the resulting region can be completely covered by trominos (each made up of three small squares of the same size as the small squares in the region) in such a way that the trominos do not overlap.*

Proof. To begin a proof by mathematical induction, first note that the theorem is certainly true for $n = 1$; the region obtained after removing a small corner square is a tromino, so it can be covered by one tromino.

Suppose that the theorem is true for $n = k$. That is, we are supposing that if a small corner square is removed from any $2^k \times 2^k$ square consisting of 2^{2k} smaller squares, then the resulting region can be covered by trominos. The proof will be established by the Principle of Mathematical Induction if we can show that the same result holds for $n = k + 1$. Consider, then, any $2^{k+1} \times 2^{k+1}$ square consisting of smaller squares. Remove one corner square to get a region that looks like this:

The region can be divided into four "medium-sized" squares, three of which are $2^k \times 2^k$ and one of which is $2^k \times 2^k$ with one corner removed, like this:

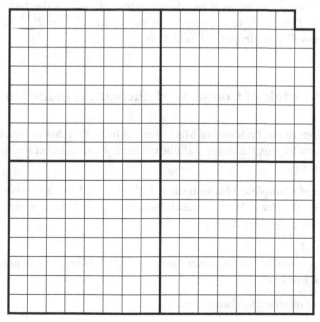

Now place a tromino in the middle of the region, as illustrated below.

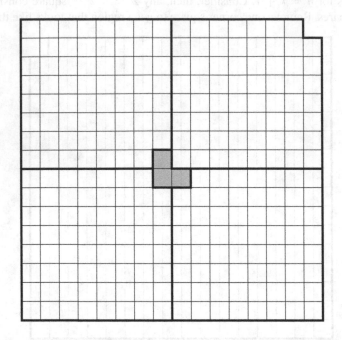

Because of the tromino in the middle, the four "medium-sized" squares remaining to be covered each have one corner covered or missing. By the inductive hypothesis, trominos can be used to cover the rest of each of the four "medium-sized" squares. This leads to a covering of the entire $2^{k+1} \times 2^{k+1}$ square, thus finishing the proof by mathematical induction.   □

2.2   The Principle of Complete Mathematical Induction

There is a variant of the Principle of Mathematical Induction that is sometimes very useful. The basis for this variant is a slightly different characterization of the set of all natural numbers.

The Principle of Complete Mathematical Induction 2.2.1. *(Sometimes   called the "Principle of Strong Mathematical Induction.") If S is any set of natural numbers with the properties that*

A. *1 is in S, and*
B. *k + 1 is in S whenever k is a natural number and all of the natural numbers from 1 through k are in S,*

then S is the set of all natural numbers.

The informal and formal proofs of the Principle of Complete Mathematical Induction are virtually the same as the proofs of the Principle of (ordinary) Mathematical Induction. First consider the informal proof. If S is any set of natural numbers with properties A and B of the Principle of Complete Mathematical Induction, then, in particular, 1 is in S. Since 1 is in S, it follows from property B that 2 is in S. Since 1 and 2 are in S, it follows from property B that 3 is in S. Since 1, 2, and 3 are in S, 4 is in S, and so on. It is suggested that you write out the details of the formal proof of the Principle of Complete Mathematical Induction as a consequence of the Well-Ordering Principle.

Just as for ordinary induction, the Principle of Complete Mathematical Induction can be generalized to begin at any natural number, not just 1.

The Generalized Principle of Complete Mathematical Induction 2.2.2. *Let m be a natural number. If S is any set of natural numbers with the properties that*

A. *m is in S, and*
B. *$k + 1$ is in S whenever k is a natural number greater than or equal to m and all of the natural numbers from m through k are in S,*

then S contains all natural numbers greater than or equal to m.

There are many situations in which it is difficult to directly apply the Principle of Mathematical Induction but easy to apply the Principle of Complete Mathematical Induction. One example of such a situation is a very precise proof of the lemma (Lemma 1.1.4) that was required to prove that there is no largest prime number.

Lemma 2.2.3. *Every natural number greater than 1 has a prime divisor.*

The following is a statement that clearly implies the above lemma. Note that we employ the convention that a single prime number is a "product of primes" where the product has only one factor.

Theorem 2.2.4. *Every natural number other than 1 is a product of prime numbers.*

Proof. We prove this theorem using the Generalized Principle of Complete Mathematical Induction starting at 2. Let S be the set of all n that are products of primes. It is clear that 2 is in S, since 2 is a prime. Suppose that every natural number from 2 up through k is in S. We must show, in order to apply the Generalized Principle of Complete Mathematical Induction, that $k + 1$ is in S.

The number $k + 1$ cannot be 1. We must therefore show that either it is prime or is a product of primes. If $k + 1$ is prime, we are done. If $k + 1$ is not prime, then $k + 1 = xy$ where each of x and y is a natural number strictly between 1 and $k + 1$. Thus x and y are each at most k, so, by the inductive hypothesis, x and y are both in S. That is, x and y are each either primes or the product of primes. Therefore, xy can be written as a product of primes by writing the product of the primes comprising x (or x itself if x is prime) times the product of the primes comprising y (or y itself if y is prime). Thus, by the Generalized Principle of Complete Mathematical Induction starting at 2, S contains all natural numbers greater than or equal to 2.   □

We now describe an interesting theorem that is a little more difficult to understand. (If you find this theorem too difficult, you need not consider it; it won't be used in anything that follows. You might wish to return to it at some later time.)

We begin by describing the case where $n = 5$. Suppose there is a pile of 5 stones. We are going to consider the sum of certain sequences of numbers obtained as follows. Begin one such sequence by dividing the pile into two smaller piles, a pile of 3 stones and a pile of 2 stones. Let the first term in the sum be $3 \cdot 2 = 6$. Repeat this process with the pile of 3 stones: divide it into a pile of 2 stones and a pile consisting of 1 stone. Add $2 \cdot 1 = 2$ to the sum. The pile with 2 stones can be divided into 2 piles of 1 stone each. Add $1 \cdot 1 = 1$ to the sum. Now go back to the pile of 2 stones created by the first division. That pile can be divided into 2 piles of 1 stone each. Add $1 \cdot 1 = 1$ to the sum. The total sum that we have is 10.

Let's create another sum in a similar manner but starting a different way. Divide the original pile of 5 stones into a pile of 4 stones and a pile of 1 stone. Begin this sum with $4 \cdot 1 = 4$. Divide the pile of 4 stones into two piles of 2 stones each and add $2 \cdot 2 = 4$ to the sum. The first pile of 2 stones can be divided into two piles of 1 stone each, so add $1 \cdot 1 = 1$ to the sum. Similarly, divide the second pile of 2 into two piles of 1 each and add $1 \cdot 1 = 1$ to the sum. The sum we get proceeding in this way is also 10.

Is it a coincidence that we got the same result, 10, for the sums we obtained in quite different ways?

Theorem 2.2.5. *For any natural number n greater than 1, consider a pile of n stones. Create a sum as follows: Divide the given pile of stones into two smaller piles. Let the product of the number of elements in one smaller pile and the number of elements in the other smaller pile be the first term in the sum. Then consider one of the smaller piles and (unless it consists of only one stone) divide that pile into two smaller piles and let the product of the number of stones in those piles be the second term in the sum. Do the same for the other smaller pile. Continue dividing, multiplying, and adding terms to the sum in all possible ways. No matter what sequence of divisions into subpiles is used, the total sum is n(n − 1)/2.*

Proof. We prove this theorem using Generalized Complete Mathematical Induction beginning with $n = 2$. Given any pile of 2 stones, there is only one way to divide it: into two piles of 1 each. Since $1 \cdot 1 = 1$, the sum is 1 in this case. Notice that $1 = 2(2 − 1)/2$, so the formula holds for the case $n = 2$.

Suppose now that the formula holds for all of $n = 2, 3, 4, \ldots, k$. Consider any pile of $k + 1$ stones. Note that $k + 1$ is at least 3. We must show that for any sequence of divisions, the resulting sum is $(k + 1)(k + 1 − 1)/2 = k(k + 1)/2$.

Begin with any division of the pile into two subpiles. Call the number of stones in the subpiles x and y respectively. Consider first the situation where $x = 1$. Then the first term in the sum is $1 \cdot y = y$. Since $x = 1$ and $x + y = k + 1$, we know that $y = k$. The process is continued by dividing the pile of y stones. By the inductive hypothesis (since $y = k$, which is greater than or equal to 2), the sum obtained by completing the process on a pile of y stones is $y(y − 1)/2$. Thus, the total sum for the original pile of $k + 1$ stones in this case is

$$y + \frac{y(y-1)}{2} = \frac{2y + (y^2 - y)}{2} = \frac{y^2 + y}{2} = \frac{y(y+1)}{2} = \frac{k(k+1)}{2}$$

If $y = 1$, the same proof can be given by simply interchanging the roles of x and y in the previous paragraph.

The last, and most interesting, case is when neither x nor y is 1. In this case, both x and y are greater than or equal to 2 and less than k. The first term in the sum is then xy. Continuing the process will give a total sum that is equal to xy plus the sum for the pile of x stones added to the sum for the pile of y stones. Therefore, using the inductive hypothesis, the sum for the original pile of $k + 1$ stones is $xy + x(x-1)/2 + y(y-1)/2$. We must show that this sum is $k(k+1)/2$.

Recall that $k + 1 = x + y$, so $x = k + 1 - y$. Using this, we see that

$$xy + \frac{x(x-1)}{2} + \frac{y(y-1)}{2}$$

$$= \frac{2(k+1-y)y}{2} + \frac{(k+1-y)(k-y)}{2} + \frac{y(y-1)}{2}$$

$$= \frac{2ky + 2y - 2y^2}{2} + \frac{k^2 + k - ky - ky - y + y^2}{2} + \frac{y^2 - y}{2}$$

$$= \frac{k^2 + k}{2}$$

$$= \frac{k(k+1)}{2}$$

This completes the proof.   □

Mathematics is the most precise of subjects. However, human beings are not always so precise; they must be careful not to make mistakes. See if you can figure out what is wrong with the "proof" of the following obviously false statement.

False Statement. All human beings are the same age.

"Proof". We will present what, at first glance at least, appears to be a proof of the above statement. We begin by reformulating it as follows: For every natural number n, every set of n people consists of people the same age. The assertion that "all human beings are the same age" would clearly follow from the case where n is the present population of the earth. We proceed by mathematical induction. The case $n = 1$ is certainly true; a set containing 1 person consists of people the same age. For the inductive step, suppose that every set of k people consists of people the same age. Let S be any set containing $k + 1$ people. We must show that all the people in S are the same age as each other.

List the people in S as follows:

$$S = \{P_1, P_2, \ldots, P_k, P_{k+1}\}$$

Consider the subset L of S consisting of the first k people in S; that is,

$$L = \{P_1, P_2, \ldots, P_k\}$$

Similarly, let R denote the subset consisting of the last k elements of S; that is,

$$R = \{P_2, \ldots, P_k, P_{k+1}\}$$

The sets L and R each contain k people, and so by the inductive hypothesis each consists of people who are the same age as each other. In particular, all the people in L are the same age as P_2. Also, all the people in R are the same age as P_2. But every person in the original set S is in either L or R, so all the people in S are the same age as P_2. Therefore, S consists of people the same age, and the assertion follows by the Principle of Mathematical Induction.

What is going on? Is it really true that all people are the same age? Not likely. Is the Principle of Mathematical Induction flawed? Or is there something wrong with the above "proof"?

Clearly there must be something wrong with the "proof." Please do not read further for at least a few minutes while you try to find the mistake.

Wait a minute. Before you read further, please try for a little bit longer to see if you can find the mistake.

If you haven't been able to find the error yourself, perhaps a hint will help. The proof of the case $n = 1$ is surely valid; a set with one person in it contains a person with whatever age that person is. What about the inductive step, going from k to $k + 1$? For it to be valid, it must apply for every natural number k. To conclude that an assertion holds for all natural numbers given that it holds for $n = 1$ requires that its truth for $n = k + 1$ is implied by its truth for $n = k$, *for every natural number k*. In fact, there is a k for which the above derivation of the case $n = k + 1$ from the case $n = k$ is not valid. Can you figure out the value of that k?

Okay, here is the mistake. Consider the inductive step when $k = 1$; that is, going from 1 to 2. In this case, the set S would have the form $S = \{P_1, P_2\}$. Then, $L = \{P_1\}$ and $R = \{P_2\}$.

The set L does consist of people the same age as each other, as does the set R. But there is no person who is in both sets. Thus, we cannot conclude that everyone in S is the same age. This shows that the above "proof" of the inductive step does not hold when $k = 1$. In fact, the following is true.

True Statement. If every pair of people in a given set of people consists of people the same age, then all the people in the set are the same age.

Proof. Let S be the given set of people; suppose $S = \{P_1, P_2, \ldots, P_n\}$. For each i from 2 to n, the pair $\{P_1, P_i\}$ consists of people the same age, by hypothesis. Thus, P_i and P_1 are the same age for every i, so every person in S is the same age as P_1. Hence, everyone in S is the same age. $\square$

2.3 Problems

Basic Exercises

1. Prove, using induction, that for every natural number n:

$$1 \cdot 2 + 2 \cdot 3 + 3 \cdot 4 + \cdots + n \cdot (n+1) = \frac{n(n+1)(n+2)}{3}$$

2. Prove, using induction, that for every natural number n:

$$\frac{1}{1 \cdot 2} + \frac{1}{2 \cdot 3} + \cdots + \frac{1}{n \cdot (n+1)} = \frac{n}{n+1}$$

3. Prove, using induction, that for every natural number n:

$$2 + 2^2 + 2^3 + \cdots + 2^n = 2^{n+1} - 2$$

4. Prove, using induction, that for every natural number n:

$$\frac{1}{2} + \frac{2}{2^2} + \frac{3}{2^3} + \cdots + \frac{n}{2^n} = 2 - \frac{n+2}{2^n}$$

Interesting Problems

5. Prove the following statement by induction: For every natural number n, every set with n elements has 2^n subsets. (A *subset* of a given set is a set all of whose elements are elements of the given set. Note that the *empty set*, the set consisting of no elements, is a subset of every set. See Section 10.1 for more information about sets.)

6. Prove, using induction, that for every natural number n:

$$1 + \frac{1}{\sqrt{2}} + \frac{1}{\sqrt{3}} + \cdots + \frac{1}{\sqrt{n}} < 2\sqrt{n}$$

7. Prove by induction that 3 divides $n^3 + 2n$, for every natural number n.
8. Show that $3^n > n^2$ for every natural number n.
9. Use induction to prove that $2^n > n^2$, for every $n > 4$.
10. Show that, for every natural number $n > 1$ and every real number r different from 1,

$$1 + r + r^2 + \cdots + r^{n-1} = \frac{r^n - 1}{r - 1}$$

Challenging Problems

11. Prove the Principle of Complete Mathematical Induction using the Well-Ordering Principle.
12. Prove the Well-Ordering Principle using the Principle of Complete Mathematical Induction.
13. One version of a game called *Nim* is played as follows. There are two players and two piles consisting of the same natural number of objects; for this example, suppose the objects are nickels. At each turn, a player removes some number of nickels from either one of the piles. Then the other player removes some number of nickels from either of the piles. The players continue playing alternately until the last nickel is removed. The winner is the player who removes the last nickel.

    Prove: If the second player always removes the same number of nickels that the first player last removed and does so from the other pile (thus making the piles equal in number after the second player's turn), then the second player will win.
14. Define the nth *Fermat number*, F_n, by $F_n = 2^{2^n} + 1$ for $n = 0, 1, 2, 3, \ldots$. The first few Fermat numbers are $F_0 = 3$, $F_1 = 5$, $F_2 = 17$, $F_3 = 257$.

    (a) Prove by induction that $F_0 \cdot F_1 \cdots F_{n-1} + 2 = F_n$, for $n \geq 1$.
    (b) Use the formula in part (a) to prove that there are an infinite number of primes, by showing that no two Fermat numbers have any prime factors in common.
        [Hint: For each F_n, let p_n be a prime divisor of F_n and show that $p_{n_1} \neq p_{n_2}$ if $n_1 \neq n_2$.]

15. The sequence of *Fibonacci numbers* is defined as follows: $x_1 = 1$, $x_2 = 1$, and, for $n > 2$, $x_n = x_{n-1} + x_{n-2}$. Prove that

$$x_n = \frac{1}{\sqrt{5}} \left[\left(\frac{1 + \sqrt{5}}{2} \right)^n - \left(\frac{1 - \sqrt{5}}{2} \right)^n \right]$$

    for every natural number n.
    [Hint: Use the fact that $x = \frac{1+\sqrt{5}}{2}$ and $x = \frac{1-\sqrt{5}}{2}$ both satisfy $1 + x = x^2$.]
16. Prove the following generalization of Theorem 2.1.6:

Theorem. *For each natural number n, consider a square consisting of 2^{2n} smaller squares (i.e., a $2^n \times 2^n$ square). If any one of the smaller squares is removed from the large square (not necessarily from the corner), then the resulting region can be completely covered by trominos (each made up of three small squares of the same size as the small squares in the region) in such a way that the trominos do not overlap.*

Chapter 3
Modular Arithmetic

Consider the number obtained by adding 3 to the number consisting of 2 to the power 3,000,005; that is, consider the number $3 + 2^{3,000,005}$. This is a very big number. Common calculators cannot deal with a number this big.

When that huge number is divided by 7, what remainder is left? You can't use your calculator because it can't count that high. However, this and similar questions are easily answered using a kind of "calculus" of divisibility and remainders that is called *modular arithmetic*. Another application of this concept will be a proof that a natural number is divisible by 9 if and only if the sum of its digits is divisible by 9. The mathematics that we develop in this chapter has numerous other applications, including, for example, providing the basis for an extremely powerful method for sending coded messages (see Chapter 6).

3.1 The Basics

Recall that we say that the integer n is *divisible* by the integer m if there exists an integer q such that $n = mq$. In this situation, we also say that m is a *divisor* of n, or m is a *factor* of n.

The definition that is fundamental for modular arithmetic is the following.

Definition 3.1.1. For any fixed natural number m greater than 1, we say that the integer *a is congruent to the integer b modulo m* if $a - b$ is divisible by m. We use the notation $a \equiv b \pmod{m}$ to denote this relationship. The number m in this notation is called the *modulus*.

Here are a few examples:

$14 \equiv 8 \pmod{3}$, since $14 - 8 = 6$ is divisible by 3
$252 \equiv 127 \pmod{5}$, since $252 - 127 = 125$ is divisible by 5
$3 \equiv -11 \pmod{7}$, since $3 - (-11) = 14$ is divisible by 7

© Springer Nature Switzerland AG 2018
D. Rosenthal et al., *A Readable Introduction to Real Mathematics*,
Undergraduate Texts in Mathematics, https://doi.org/10.1007/978-3-030-00632-7_3

Congruence shares an important property with equality.

Theorem 3.1.2. *If $a \equiv b$ (mod m) and $b \equiv c$ (mod m), then $a \equiv c$ (mod m).*

Proof. The hypothesis states that $a - b$ and $b - c$ are both divisible by m; that is, there are integers t and s such that $a - b = tm$ and $b - c = sm$. Thus, $a - c = a - b + b - c = tm + sm = (t + s)m$. In other words, $a - c$ is divisible by m. By definition, then, $a \equiv c$ (mod m).   □

The theorem just proven shows that we can replace numbers in a congruence modulo m by any numbers congruent to them modulo m.

Although the modulus m must be bigger than 1, there is no such restriction on the integers a and b; they could even be negative. In the case where a and b are positive integers, the relationship $a \equiv b$ (mod m) can be expressed in more familiar terms.

Theorem 3.1.3. *When a and b are nonnegative integers, the relationship $a \equiv b$ (mod m) is equivalent to a and b leaving equal remainders upon division by m.*

Proof. Consider dividing m into a; if it "goes in evenly," then m is a divisor of a and the remainder r is 0. In any case, there are nonnegative integers q and r such that $a = qm + r$; q is the quotient and r is the remainder. The nonnegative number r is less than m, since it is the remainder. Similarly, divide b by m, getting $b = q_0 m + r_0$ for some q_0 and some r_0. This yields

$$a - b = (qm + r) - (q_0 m + r_0) = m(q - q_0) + (r - r_0)$$

If $r = r_0$, then $a - b$ is obviously divisible by m, so $a \equiv b$ (mod m). Conversely, if r is not equal to r_0, note that $r - r_0$ cannot be a multiple of m. (This follows from the fact that r and r_0 are both nonnegative numbers which are strictly less than m.) Thus, $a - b$ is a multiple of m plus a number that is not a multiple of m, and therefore $a - b$ is not a multiple of m. That is, it is not the case that $a \equiv b$ (mod m).   □

A special case of the above theorem is that a positive number is congruent modulo m to the remainder it leaves upon division by m. The possible remainders that arise from division by a given natural number m are $0, 1, 2, \ldots, m - 1$.

Theorem 3.1.4. *For a given modulus m, each integer is congruent to exactly one of the numbers in the set $\{0, 1, 2, \ldots, m - 1\}$.*

Proof. Let a be an integer. If a is positive, the result follows from the fact, discussed above, that a is congruent to the remainder it leaves upon division by m. If a is not positive, choosing t big enough would make $tm + a$ positive. For such a t, $tm + a$ is congruent to the remainder it leaves upon division by m. But also $tm + a \equiv a$ (mod m). It follows from Theorem 3.1.2 that a is congruent to the remainder that $tm + a$ leaves upon division by m. An integer cannot be congruent to two different numbers in the given set $\{0, 1, 2, \ldots, m - 1\}$, since no two numbers in the set are congruent to each other (by Theorem 3.1.3).   □

For a fixed modulus, congruences have some properties that are similar to those for equations.

Theorem 3.1.5. *If $a \equiv b$ (mod m) and $c \equiv d$ (mod m), then*

(i) $(a + c) \equiv (b + d)$ (mod m), *and*
(ii) $ac \equiv bd$ (mod m).

Proof. To prove (i), note that $a \equiv b$ (mod m) means that $a - b = sm$ for some integer s. Similarly, $c - d = tm$ for some integer t. The conclusion we are trying to establish is equivalent to the assertion that $(a + c) - (b + d)$ is a multiple of m. But $(a + c) - (b + d) = (a - b) + (c - d)$, which is equal to $sm + tm = (s + t)m$, so the result follows.

To prove (ii), note that from $a - b = sm$ and $c - d = tm$, we get $a = b + sm$ and $c = d + tm$, so

$$ac = (b + sm)(d + tm) = bd + btm + smd + stm^2$$

It follows that $ac - bd = m(bt + sd + stm)$, so $ac - bd$ is a multiple of m and the result is established. $\square$

Theorem 3.1.5 tells us that congruences are similar to equations in that you can add congruent numbers to both sides of a congruence or multiply both sides of a congruence by congruent numbers and preserve the congruence, as long as all the congruences are with respect to the same fixed modulus.

For example, since $3 \equiv 28$ (mod 5) and $17 \equiv 2$ (mod 5), it follows that $20 \equiv 30$ (mod 5) and $51 \equiv 56$ (mod 5).

Here is another example: $8 \equiv 1$ (mod 7), so $8^2 \equiv 1^2$ (mod 7), or $8^2 \equiv 1$ (mod 7). It follows that $8^2 \cdot 8 \equiv 1 \cdot 1$ (mod 7), or $8^3 \equiv 1$ (mod 7). In fact, all positive integer powers of 8 are congruent to 1 modulo 7. This is a special case of the next result.

Theorem 3.1.6. *If $a \equiv b$ (mod m), then $a^n \equiv b^n$ (mod m), for every natural number n.*

Proof. We use the Principle of Mathematical Induction. The case $n = 1$ is the hypothesis. Assume that the result is true for $n = k$; that is, $a^k \equiv b^k$ (mod m). Since $a \equiv b$ (mod m), part (ii) of Theorem 3.1.5 gives $a \cdot a^k \equiv b \cdot b^k$ (mod m), or $a^{k+1} \equiv b^{k+1}$ (mod m). Thus, the statement is true for all $n \geq 1$ by induction. $\square$

3.2 Some Applications

We can use the above to easily solve the problem that we mentioned at the beginning of this chapter: What is the remainder left when $3 + 2^{3,000,005}$ is divided by 7?

First note that $2^3 = 8$ is congruent to 1 modulo 7. Therefore, by Theorem 3.1.6, $(2^3)^{1,000,000}$ is congruent to $1^{1,000,000}$ modulo 7. But $1^{1,000,000} = 1$. Thus $2^{3,000,000} \equiv 1$ (mod 7). Since $2^5 \equiv 4$ (mod 7) and $2^{3,000,005} = 2^{3,000,000} \cdot 2^5$, it follows that $2^{3,000,005} \equiv 4$ (mod 7). Thus, $3 + 2^{3,000,005} \equiv 3 + 4 \equiv 0$ (mod 7).

Therefore, 7 is a divisor of $3 + 2^{3,000,005}$. In other words, the remainder that is left when $3 + 2^{3,000,005}$ is divided by 7 is 0.

Let's look at the next question we mentioned at the beginning of this chapter, the relationship between divisibility by 9 of a number and divisibility by 9 of the sum of the digits of the number. To illustrate, we begin with a particular example. Consider the number 73,486. What that really means is

$$7 \cdot 10^4 + 3 \cdot 10^3 + 4 \cdot 10^2 + 8 \cdot 10 + 6$$

Note that 10 is congruent to 1 modulo 9, so 10^n is congruent to 1 modulo 9 for every natural number n. Thus, $a \cdot 10^n \equiv a$ (mod 9) for every a and every n. It follows that $7 \cdot 10^4 + 3 \cdot 10^3 + 4 \cdot 10^2 + 8 \cdot 10 + 6$ is congruent to $(7+3+4+8+6)$ modulo 9. Thus, the number 73,486 and the sum of its digits are congruent to each other modulo 9 and therefore leave the same remainders upon division by 9. The general theorem is the following.

Theorem 3.2.1. *Every natural number is congruent to the sum of its digits modulo 9. In particular, a natural number is divisible by 9 if and only if the sum of its digits is divisible by 9.*

Proof. If n is a natural number, then we can write it in terms of its digits in the form $a_k a_{k-1} a_{k-2} \ldots a_1 a_0$ (note that this is a listing of digits, not a product of digits), where each a_i is one of 0, 1, 2, 3, 4, 5, 6, 7, 8, 9 (with $a_k \neq 0$). That is, a_0 is the digit in the "1's place," a_1 is the digit in the "10's place," a_2 is the digit in the "100's place," and so on. (In the previous example, n was the number 73,486, so in that case $a_4 = 7, a_3 = 3, a_2 = 4, a_1 = 8$, and $a_0 = 6$.) This means that

$$n = a_k \cdot 10^k + a_{k-1} \cdot 10^{k-1} + a_{k-2} \cdot 10^{k-2} + \cdots + a_2 \cdot 10^2 + a_1 \cdot 10 + a_0$$

As shown above, $10 \equiv 1$ (mod 9) implies $10^i \equiv 1$ (mod 9), for every positive integer i. Therefore, n is congruent to $(a_k + a_{k-1} + a_{k-2} + \cdots + a_1 + a_0)$ modulo 9. Thus, n and the sum of its digits leave the same remainders upon division by 9. In particular, n is divisible by 9 if and only if the sum of its digits is divisible by 9. $\square$

Congruences with unknowns can easily be solved by just trying all possibilities if the modulus is small.

Example 3.2.2. Find a solution to the congruence $5x \equiv 11$ (mod 19).

Solution. If there is a solution, then, by Theorem 3.1.4, there is a solution within the set $\{0, 1, 2, \ldots, 18\}$. If $x = 0$, then $5x = 0$, so 0 is not a solution. Similarly, for $x = 1, 5x = 5$; for $x = 2, 5x = 10$; for $x = 3, 5x = 15$; and for $x = 4, 5x = 20$. None of these are congruent to 11 (mod 19), so we have not yet found a solution. However, when $x = 6, 5x = 30$, which is congruent to 11 (mod 19). Thus, $x \equiv 6$ (mod 19) is a solution of the congruence. $\square$

Example 3.2.3. Show that there is no solution to the congruence $x^2 \equiv 3$ (mod 5).

Proof. If $x = 0$, then $x^2 = 0$; if $x = 1$, then $x^2 = 1$; if $x = 2$, then $x^2 = 4$; if $x = 3$, then $x^2 = 9$, which is congruent to 4 (mod 5); and if $x = 4$, then $x^2 = 16$ which is congruent to 1 (mod 5). If there was any solution, it would be congruent to one of $\{0, 1, 2, 3, 4\}$ by Theorem 3.1.4. Thus, the congruence has no solution. ◻

3.3 Problems

Basic Exercises

1. Find a solution x to each of the following congruences. ("Solution" means integer solution.)

   (a) $2x \equiv 7$ (mod 11)
   (b) $7x \equiv 4$ (mod 11)
   (c) $x^5 \equiv 3$ (mod 4)

2. For each of the following congruences, either find a solution or prove that no solution exists.

   (a) $39x \equiv 13$ (mod 5)
   (b) $95x \equiv 13$ (mod 5)
   (c) $x^2 \equiv 3$ (mod 6)
   (d) $5x^2 \equiv 12$ (mod 8)
   (e) $4x^3 + 2x \equiv 7$ (mod 5)

Interesting Problems

3. Find the remainder when:

   (a) 3^{2463} is divided by 8
   (b) 2^{923} is divided by 15
   (c) 243^{101} is divided by 8
   (d) $5^{2001} + (27)!$ is divided by 8
   (e) $(-8)^{4124} + 6^{3101} + 7^5$ is divided by 3
   (f) $7^{103} + 6^{5409}$ is divided by 3
   (g) $5! \cdot 181 - 866 \cdot 332$ is divided by 6

4. Is $2^{598} + 3$ divisible by 15?
5. Find a digit b such that the number $2794b2$ is divisible by 8.
6. Determine whether or not $17^{2492} + 25^{376} + 5^{782}$ is divisible by 3.
7. Suppose that 7^{22} is written out in the ordinary way. What is its last digit?

8. Determine whether or not the following congruence has a natural number solution:

$$5^x + 3 \equiv 5 \pmod{100}$$

9. Prove that $n^2 - 1$ is divisible by 8, for every odd integer n.

10. Prove that a natural number is divisible by 3 if and only if the sum of its digits is divisible by 3.

11. Prove that $x^5 \equiv x \pmod{10}$, for every integer x. (This shows that x^5 and x have the same units' digit for every integer x.)

12. Suppose a number is written as $abba$, where a and b are any integers from 1 to 9. Prove that this number is divisible by 11.

13. Find the units' digit of 27493^{6792}.

14. Show that if m is a natural number and a is a negative integer, then there exists an r with $0 \leq r \leq m - 1$ and an integer q such that $a = qm + r$. (See the proof of Theorem 3.1.3.)

15. Prove that, for every pair of natural numbers m and n, m^2 is congruent to n^2 modulo $(m + n)$.

Challenging Problems

16. Prove that 5 divides $3^{2n+1} + 2^{2n+1}$, for every natural number n.

17. Prove that 7 divides $8^{2n+1} + 6^{2n+1}$, for every natural number n.

18. Prove that a natural number that is congruent to 2 modulo 3 has a prime factor that is congruent to 2 modulo 3.

19. If m is a natural number greater than 1 and is not prime, then we know that $m = ab$, where $1 < a < m$ and $1 < b < m$. Show that there is no integer x such that $ax \equiv 1 \pmod{m}$. (That is, a has no *multiplicative inverse modulo m*. The situation is different if m is prime: see Problem 7 in Chapter 4.)

20. Prove that 133 divides $11^{n+1} + 12^{2n-1}$, for every natural number n.

21. A natural number r less than or equal to $m - 1$ is called a *quadratic residue modulo m* if there is an integer x such that $x^2 \equiv r \pmod{m}$. Determine all the quadratic residues modulo 11.

22. Show that there do not exist natural numbers x and y such that $x^2 + y^2 = 4003$. [Hint: Begin by determining which of the numbers $\{0, 1, 2, 3\}$ can be congruent to $x^2 \pmod{4}$.]

23. Discover and prove a theorem determining whether a natural number is divisible by 11, in terms of its digits.

24. Prove that there are an infinite number of primes of the form $4k + 3$ with k a natural number.
[Hint: If $p_1, p_2, \ldots, p_n$ are n such primes, show that $(4 \cdot p_1 \cdot p_2 \cdots p_n) - 1$ has at least one prime divisor of the given form.]

25. Prove that there are an infinite number of primes of the form $6k + 5$ with k a natural number.
26. Prove that every prime number greater than 3 differs by 1 from a multiple of 6.
27. Show that, if x, y, and z are integers such that $x^2 + y^2 = z^2$, then at least one of $\{x, y, z\}$ is divisible by 2, at least one of $\{x, y, z\}$ is divisible by 3, and at least one of $\{x, y, z\}$ is divisible by 5.
28. Let $p(x)$ be a non-constant polynomial with integer coefficients. (That is, there exists a natural number n and integers a_i such that $p(x) = a_n x^n + a_{n-1} x^{n-1} + \cdots + a_1 x + a_0$ and $a_n \neq 0$.) Let a, k, and m be integers with $m > 1$. Suppose that $p(a) \equiv k \pmod{m}$. Prove that $p(a + m) \equiv k \pmod{m}$.
29. Show that the polynomial $p(x) = x^2 - x + 41$ takes prime values for every x in the set $\{0, 1, 2, \ldots, 40\}$.
30. Show that there does not exist any non-constant polynomial $p(x)$ with integer coefficients such that $p(x)$ is a prime number for all natural numbers x.

Chapter 4
The Fundamental Theorem of Arithmetic

Is $13^{217} \cdot 37^{92} \cdot 41^{15} = 19^{111} \cdot 29^{145} \cdot 43^{12} \cdot 47^5$?

We have seen that every natural number greater than 1 is either a prime or a product of primes. The above equation, if it was an equation, would express a number in two different ways as a product of primes. Does the representation of a natural number as a product of primes have to be unique? The answer is obviously "no" in one sense. For example, $6 = 3 \cdot 2 = 2 \cdot 3$. Thus, the same number can be written in two different ways as a product of primes if we consider different orders as "different ways." But suppose that we don't consider the ordering; must the factorization of a natural number into a product of primes be unique except for the order? For example, could the above equation hold?

In fact, every natural number other than 1 has a factorization into a product of primes and the factorization is unique except for the order. This result is so important that it is called the *Fundamental Theorem of Arithmetic*. We will give two proofs. The second proof requires a little more development and will be given later (Theorem 7.2.4). The first proof is short but tricky.

4.1 Proof of the Fundamental Theorem of Arithmetic

In order to simplify the statement of the Fundamental Theorem of Arithmetic, we use the expression "a product of primes" to include the case of a single prime number (as we did in Theorem 2.2.4).

The Fundamental Theorem of Arithmetic 4.1.1. *Every natural number greater than 1 can be written as a product of primes, and the expression of a number as a product of primes is unique except for the order of the factors.*

Proof. We have already established that every natural number greater than 1 can be written as a product of primes (see Theorem 2.2.4). That is the easy part of the

© Springer Nature Switzerland AG 2018

D. Rosenthal et al., *A Readable Introduction to Real Mathematics*,
Undergraduate Texts in Mathematics, https://doi.org/10.1007/978-3-030-00632-7_4

Fundamental Theorem of Arithmetic; the harder part is the uniqueness. The proof of uniqueness that we present below is a proof by contradiction. That is, we will assume that there is a natural number with more than one representation as a product of primes and derive a contradiction from this assumption, thereby showing that this assumption is incorrect.

Suppose, then, that there is at least one natural number greater than 1 with at least two different representations as a product of primes. By the Well-Ordering Principle (2.1.2), there would then be a smallest natural number with that property (i.e., a smallest natural number that has at least two different such representations). Let N be that smallest such number. Write out two different factorizations of N:

$$N = p_1 p_2 \cdots p_r = q_1 q_2 \cdots q_s$$

where each of the p_i and the q_j are primes (there can be repetitions of the same prime). Notice that N cannot be prime, since there is only one way to express a prime, so r and s are both bigger than 1.

We first claim that no p_i could be equal to any q_j. This follows from the fact that N is the smallest number with a non-unique representation, for if $p_i = q_j$ for some i and j, that common factor could be divided from both of the two different factorizations for N, producing a smaller number that has at least two different factorizations. Thus, no p_i is equal to any q_j.

Since p_1 is different from q_1, one of p_1 and q_1 is less than the other; suppose that p_1 is less than q_1. (If q_1 is less than p_1, the same proof could be repeated by simply interchanging the p's and q's.) Define M by

$$M = N - (p_1 q_2 \cdots q_s)$$

Then M is a natural number that is less than N. Substituting the product $p_1 p_2 \cdots p_r$ for N gives

$$M = (p_1 p_2 \cdots p_r) - (p_1 q_2 \cdots q_s) = p_1 [(p_2 \cdots p_r) - (q_2 \cdots q_s)]$$

from which it follows that p_1 divides M. In particular, M is not 1. Since M is less than N, M has a unique factorization into primes.

Substituting the product $q_1 q_2 \cdots q_s$ for N in the definition of M gives a different expression:

$$M = (q_1 q_2 \cdots q_s) - (p_1 q_2 \cdots q_s) = (q_1 - p_1)(q_2 \cdots q_s)$$

The unique factorization of M into primes can thus be obtained by writing the unique factorization of $q_1 - p_1$ followed by the product $q_2 \cdots q_s$. On the other hand, the fact that p_1 is a divisor of M implies that p_1 must appear in the factorization of M into primes. Since p_1 is distinct from each of $\{q_2, \ldots, q_s\}$, it follows that p_1 must occur in the factorization of $q_1 - p_1$ into primes. Thus, $q_1 - p_1 = p_1 k$, for some natural number k. It follows that $q_1 = p_1 + p_1 k = p_1(1 + k)$, which shows that q_1

is divisible by p_1. Since p_1 and q_1 are distinct primes, this is impossible. Hence, the assumption that there is a natural number with two distinct factorizations leads to a contradiction, so factorizations into primes are unique.   □

The Fundamental Theorem of Arithmetic gives a so-called "canonical form" for expressing each natural number greater than 1.

Corollary 4.1.2. *Every natural number N greater than 1 has a canonical factorization into primes; that is, each natural number N greater than 1 has a unique representation of the form $N = p_1^{\alpha_1} p_2^{\alpha_2} \cdots p_n^{\alpha_n}$, where each p_i is a prime, p_i is less than p_{i+1} for each i, and each α_i is a natural number.*

Proof. To see this, simply factor the given number as a product of primes and then collect all occurrences of the smallest prime together, then all the occurrences of the next smallest prime, and so on.   □

For example, the canonical form of 60,368 is $2^4 \cdot 7^3 \cdot 11$. The canonical form of 19 is simply 19.

As we will see, the following corollary of the Fundamental Theorem of Arithmetic is very useful. (If the corollary below is independently established, then it is easy to derive the Fundamental Theorem of Arithmetic from it. In fact, most presentations of the proof of the Fundamental Theorem of Arithmetic use this approach rather than the shorter but trickier proof that we gave above. We will present such a proof later (Theorem 7.2.4).)

Corollary 4.1.3. *If p is a prime number and a and b are natural numbers such that p divides ab, then p divides at least one of a and b. (That is, if a prime divides a product, then it divides at least one of the factors.)*

Proof. Since p divides ab, there is some natural number d such that $ab = pd$. The unique factorization of ab into primes therefore contains the prime p and all the primes that divide d. On the other hand, a and b each have unique factorizations into primes. Let the canonical factorization of a be $q_1^{\alpha_1} q_2^{\alpha_2} \cdots q_m^{\alpha_m}$ and of b be $r_1^{\beta_1} r_2^{\beta_2} \cdots r_n^{\beta_n}$. Then,

$$ab = (q_1^{\alpha_1} q_2^{\alpha_2} \cdots q_m^{\alpha_m})(r_1^{\beta_1} r_2^{\beta_2} \cdots r_n^{\beta_n})$$

Since the factorization of ab into primes is unique, p must occur either as one of the q_i's, in which case p divides a, or as one of the r_j's, in which case p divides b. Thus, p divides at least one of a and b, and the corollary is established.   □

It should be noted that this corollary does not generally hold for divisors that are not prime. For example, 18 divides $3 \cdot 12$, but 18 does not divide 3 and 18 does not divide 12.

4.2   Problems

Basic Exercises

1. Find the canonical factorization into primes of each of the following:

   (a) 52   (e) $122 \cdot 54$
   (b) 72   (f) 112
   (c) 47   (g) 224
   (d) 625   (h) $112 + 224$

2. Find natural numbers x, y, and z such that

   (a) $3^x \cdot 100 \cdot 5^y = 9 \cdot 10^z \cdot 5$
   (b) $50 \cdot 2^y \cdot 7^z = 5^x \cdot 2^3 \cdot 14$

3. Show that if p is a prime number and $a_1, a_2, \ldots, a_n$ are natural numbers such that p divides the product $a_1 a_2 \cdots a_n$, then p divides a_i for at least one a_i.
4. Show that if p is a prime number and a and n are natural numbers such that p divides a^n, then p divides a.

Interesting Problems

5. Find the smallest natural numbers x and y such that

   (a) $7^2 x = 5^3 y$
   (b) $2^5 x = 10^2 y$
   (c) $127x = 54y$

6. Find nonnegative integers w, x, y, and z such that

$$17^2 25^2 2^z = 10^x 34^y 7^w$$

Challenging Problems

7. Suppose that p is a prime number and p does not divide a. Prove that the congruence $ax \equiv 1 \pmod{p}$ has a solution. (This proves that a has a *multiplicative inverse modulo* p.)
8. Prove that a natural number m greater than 1 is prime if m has the property that it divides at least one of a and b whenever it divides ab.
9. Prove that, for every prime number p, $x^2 \equiv 1 \pmod{p}$ implies $x \equiv 1 \pmod{p}$ or $x \equiv (p - 1) \pmod{p}$.

10. Suppose that a and b are natural numbers whose prime factorizations have no primes in common (the pair a, b is then said to be *relatively prime*; see Definition 7.2.1). Show that, for any natural number m, the product ab divides m if each of a and b divides m.

11. Using the result of Problem 10:

    (a) Prove that 42 divides $n^7 - n$, for every natural number n.

    (b) Prove that 21 divides $3n^7 + 7n^3 + 11n$, for every natural number n.

Chapter 5
Fermat's Little Theorem and Wilson's Theorem

We've seen that we can add or multiply "both sides" of a congruence by congruent numbers and the result will be a congruence (Theorem 3.1.5). What about dividing both sides of a congruence by the same natural number? For the result to have a chance of being a congruence, the divisor must divide evenly into both sides of the congruence so that the result involves only integers, not fractions (congruences are only defined for integers). However, even that condition is not sufficient to ensure that the result will be a congruence. For example, $6 \cdot 2$ is congruent to $6 \cdot 1$ modulo 3, but 2 is not congruent to 1 modulo 3. This is not a surprising example, since 6 is congruent to 0 modulo 3, so "dividing both sides" of the above congruence by 6 is like dividing by 0, which gives wrong results for equations as well. However, there are also examples where dividing both sides of a congruence by a number that is not congruent to 0 leads to results that are not congruent. For example, $12 \cdot 3$ is congruent to $24 \cdot 3$ modulo 9, but 12 is not congruent to 24 modulo 9, in spite of the fact that 3 is not congruent to 0 modulo 9.

However, there are important cases in which we can divide both sides of a congruence and be assured that the result is a congruence. Analyzing these cases produces some very interesting theorems.

5.1 Fermat's Little Theorem

Theorem 5.1.1. *If p is a prime and a is not divisible by p, and if $ab \equiv ac$ (mod p), then $b \equiv c$ (mod p). (That is, we can divide both sides of a congruence modulo a prime by any natural number that divides both sides of the congruence and is not divisible by the prime.)*

Proof. We are given that p divides $ab - ac$. This is the same as saying that p divides $a(b - c)$. Corollary 4.1.3 shows that, since p divides $a(b - c)$, p must also divide

© Springer Nature Switzerland AG 2018
D. Rosenthal et al., *A Readable Introduction to Real Mathematics*,
Undergraduate Texts in Mathematics, https://doi.org/10.1007/978-3-030-00632-7_5

either a or $b - c$. Since the hypothesis states that a is not divisible by p, this implies that $b - c$ must be divisible by p. That is the same as saying $b \equiv c$ (mod p).   □

Consider any given prime number p. The possible remainders when a natural number is divided by p are the numbers $\{0, 1, \ldots, p - 1\}$. By Theorem 3.1.4, no two of these numbers are congruent to each other and every natural number (in fact, every integer) is congruent modulo p to one of those numbers. An integer is divisible by p if and only if it is congruent to 0 modulo p. Thus, each integer that is not divisible by p is congruent to exactly one of the numbers in the set $\{1, 2, \ldots, p-1\}$. This is the basis for the proof of the following beautiful, and very useful, theorem.

Fermat's Little Theorem 5.1.2. *If p is a prime number and a is any natural number that is not divisible by p, then $a^{p-1} \equiv 1$ (mod p).*

Proof. Let p be any prime number and let a be any natural number that is not divisible by p. Consider the set of numbers $\{a \cdot 1, a \cdot 2, \ldots, a \cdot (p - 1)\}$. First note that no two of those numbers are congruent to each other, for if $as \equiv at$ (mod p), then, by Theorem 5.1.1, $s \equiv t$ (mod p). Since no two of the numbers in the set $\{1, 2, \ldots, p - 1\}$ are congruent to each other, this shows that the same is true of the numbers in the set $\{a \cdot 1, a \cdot 2, \ldots, a \cdot (p - 1)\}$. Also note that each of the numbers in the set $\{a \cdot 1, a \cdot 2, \ldots, a \cdot (p - 1)\}$ is congruent to one of the numbers in $\{1, 2, \ldots, p - 1\}$ since no number in either set is divisible by p. Thus, the numbers in the set $\{a \cdot 1, a \cdot 2, \ldots, a \cdot (p - 1)\}$ are congruent, in some order, to the numbers in the set $\{1, 2, \ldots, p - 1\}$. This implies that the product of all of the numbers in the set $\{a \cdot 1, a \cdot 2, \ldots, a \cdot (p - 1)\}$ is congruent modulo p to the product of all the numbers in $\{1, 2, \ldots, p - 1\}$. Thus, $a \cdot 1 \cdot a \cdot 2 \cdots a \cdot (p - 1)$ is congruent to $1 \cdot 2 \cdot 3 \cdots (p - 1)$ modulo p. Since the number a occurs $p - 1$ times in this congruence, this yields $a^{p-1}(1 \cdot 2 \cdot 3 \cdots (p - 1)) \equiv (1 \cdot 2 \cdot 3 \cdots (p - 1))$ (mod p). Clearly, p does not divide $1 \cdot 2 \cdot 3 \cdots (p - 1)$ (by repeated application of Corollary 4.1.3). Thus, by Theorem 5.1.1, we can "divide" both sides of the above congruence by $1 \cdot 2 \cdot 3 \cdots (p - 1)$, yielding $a^{p-1} \equiv 1$ (mod p).   □

As we shall see, Fermat's Little Theorem has important applications, including in establishing a method for sending coded messages. It is also sometimes useful to apply Fermat's Little Theorem to specific cases. For example, $88^{100} - 1$ is divisible by 101. (Don't try to verify this on your calculator!)

The following corollary of Fermat's Little Theorem is sometimes useful since it doesn't require that a not be divisible by p.

Corollary 5.1.3. *If p is a prime number, then $a^p \equiv a$ (mod p) for every natural number a.*

Proof. If p does not divide a, then Fermat's Little Theorem states that $a^{p-1} \equiv 1$ (mod p). Multiplying both sides of this congruence by a gives the result in this case. On the other hand, if p does divide a, then p also divides a^p, so a^p and a are both congruent to 0 mod p.   □

Definition 5.1.4. A *multiplicative inverse modulo* p for a natural number a is a natural number b such that $ab \equiv 1 \pmod{p}$.

Of course, if b is a multiplicative inverse of a modulo p, then so is any natural number that is congruent to b modulo p.

Fermat's Little Theorem provides one way of showing that all natural numbers that are not multiples of a given prime p have multiplicative inverses modulo p.

Corollary 5.1.5. *If p is a prime and a is a natural number that is not divisible by p, then there exists a natural number x such that $ax \equiv 1 \pmod{p}$.*

Proof. In the case where $p = 2$, each such a must be congruent to 1 modulo 2, so we can take $x = 1$. If p is greater than 2, then, for each given a, let $x = a^{p-2}$. Then $ax = a \cdot a^{p-2} = a^{p-1}$ and, by Fermat's Little Theorem, $a^{p-1} \equiv 1 \pmod{p}$.   □

It turns out to be interesting and useful to know which natural numbers are congruent to their own inverses modulo p. If x is such a number, then $x \cdot x \equiv 1 \pmod{p}$. In other words, such an x is a solution to the congruence $x^2 \equiv 1 \pmod{p}$, or $x^2 - 1 \equiv 0 \pmod{p}$. The solutions of the equation $x^2 - 1 = 0$ are $x = 1$ and $x = -1$. The solutions of the congruence are similar.

Theorem 5.1.6. *If p is a prime number and x is an integer satisfying $x^2 \equiv 1 \pmod{p}$, then either $x \equiv 1 \pmod{p}$ or $x \equiv p-1 \pmod{p}$. (Note that $p-1 \equiv -1 \pmod{p}$.)*

Proof. If $x^2 \equiv 1 \pmod{p}$, then, by definition, p divides $x^2 - 1$. But $x^2 - 1 = (x - 1)(x + 1)$. Since p divides $x^2 - 1$, Corollary 4.1.3 implies that p divides at least one of $x - 1$ and $x + 1$. If p divides $x - 1$, then $x \equiv 1 \pmod{p}$. If p divides $x + 1$, then $x \equiv -1 \pmod{p}$, or $x \equiv p - 1 \pmod{p}$.   □

The following lemma is needed in the proof of Wilson's Theorem (5.2.1).

Lemma 5.1.7. *If a and c have a common multiplicative inverse modulo p, then a is congruent to c modulo p.*

Proof. Suppose $ab \equiv 1 \pmod{p}$ and $cb \equiv 1 \pmod{p}$. Then multiplying the second congruence on the right by a yields $cba \equiv a \pmod{p}$ and, since $ba \equiv 1 \pmod{p}$, this gives $c \equiv a \pmod{p}$.   □

5.2 Wilson's Theorem

As we now show, these considerations lead to a proof of Wilson's Theorem, a theorem that is very beautiful, although it is considerably less famous and much less useful than Fermat's Little Theorem.

Wilson's Theorem 5.2.1. *If p is a prime number, then $(p - 1)! + 1 \equiv 0 \pmod{p}$. (In other words, if p is prime, then p divides $(p - 1)! + 1$.)*

Proof. First note that the theorem is obviously true when $p=2$; in this case, it states $(1 + 1) \equiv 0$ (mod 2). In the following, we assume that p is a prime greater than 2.

As we indicated above, a multiplicative inverse of an integer x modulo p is an integer y such that $xy \equiv 1$ (mod p). As we have seen, every number in the set $\{1, 2, \ldots, p - 1\}$ is distinct modulo p, and, by Corollary 5.1.5, each has a multiplicative inverse modulo p. Since no multiplicative inverse can be divisible by p, the multiplicative inverse of each number in $\{1, 2, \ldots, p - 1\}$ is congruent to one of the numbers in $\{1, 2, \ldots, p - 1\}$. By Theorem 5.1.6, the only numbers in the set $\{1, 2, \ldots, p - 1\}$ that are congruent to their own multiplicative inverses are the numbers 1 and $p - 1$. Leave those two numbers aside for the moment. Note that if y is a multiplicative inverse of x, then x is a multiplicative inverse of y. Thus, the numbers in the set $\{2, 3, \ldots, p - 2\}$ each have multiplicative inverses in that same set, and each number in that set differs from its multiplicative inverse. By Lemma 5.1.7, no two numbers in the set can have the same inverse. Therefore, we can arrange the numbers in the set $\{2, 3, \ldots, p - 2\}$ in pairs consisting of a number and its multiplicative inverse. Since the product of a number and its multiplicative inverse is congruent to 1 modulo p, the product of all the numbers in the set $\{2, 3, \ldots, p - 2\}$ is congruent to 1 modulo p. Thus, $2 \cdot 3 \cdots (p - 2) \equiv 1$ (mod p). Multiplying both sides by 1 gives $1 \cdot 2 \cdot 3 \cdots (p - 2) \equiv 1$ (mod p). Now $p - 1 \equiv -1$ (mod p), so $1 \cdot 2 \cdots (p - 2) \cdot (p - 1) \equiv 1 \cdot (-1)$ (mod p). In other words, $(p - 1)! \equiv -1$ (mod p), which yields $(p - 1)! + 1 \equiv 0$ (mod p).   □

Theorem 5.2.2. *If m is a composite number larger than 4, then $(m - 1)! \equiv 0$ (mod m) (so that $(m - 1)! + 1 \equiv 1$ (mod m)).*

Proof. Let m be any composite number larger than 4. We must show that $(m - 1)!$ is divisible by m. If $m = ab$, with a different from b and both less than m, then a and b each occur as distinct factors in $(m - 1)!$. Thus, $m = ab$ is a factor of $(m - 1)!$, so $(m - 1)!$ is congruent to 0 modulo m. The only composite numbers m that cannot be written as a product of two distinct natural numbers less than m are those numbers that are squares of primes. (To see this, use the fact that every composite can be written as a product of primes.) Thus, the only remaining case to prove is when $m = p^2$ for some prime p. In this case, if m is larger than 4, then p is a prime bigger than 2. In that case, p^2 is greater than $2p$. Thus, $p^2 - 1$ is greater than or equal to $2p$, so $(p^2 - 1)!$ contains the factor $2p$ as well as the factor p. Therefore, $(p^2 - 1)!$ contains the product $2p^2$. In particular, $(m - 1)!$ is divisible by $m = p^2$.   □

The following combines Wilson's Theorem and its converse.

Theorem 5.2.3. *If m is a natural number other than 1, then $(m - 1)! + 1 \equiv 0$ (mod m) if and only if m is a prime number.*

Proof. This follows immediately from Wilson's Theorem when m is prime and from the previous theorem for composite m in all cases except for $m = 4$. If $m = 4$, then $(m - 1)! + 1 = 3! + 1 = 7$, which is not congruent to 0 modulo 4, so the theorem holds for all m.   □

It might be thought that Wilson's Theorem would provide a good way to check whether or not a given number m is prime: simply see whether m divides $(m-1)!+1$. However, the fact that $(m-1)!$ is so much larger than m makes this a very impractical way of testing primality for large values of m.

5.3 Problems

Basic Exercises

1. Find the remainder when 24^{103} is divided by 103.
2. Find a solution x to each of the following congruences:

   (a) $2^x \equiv 1 \pmod{103}$
   (b) $16! \cdot x \equiv 5 \pmod{17}$

3. Find the remainder when $99^{100} - 1$ is divided by 101.

Interesting Problems

4. Suppose that p is a prime greater than 2 and $a \equiv b^2 \pmod p$ for some natural number b that is not divisible by p. Prove that $a^{\frac{p-1}{2}} \equiv 1 \pmod p$.
5. Find three different prime factors of $10^{12} - 1$.
6. Let p be a prime number. Prove that $1^2 \cdot 2^2 \cdot 3^2 \cdots (p-1)^2 - 1$ is divisible by p.
7. For each of the following congruences, either find a solution or prove that no solution exists.

   (a) $102! \cdot x + x \equiv 4 \pmod{103}$
   (b) $x^{16} - 2 \equiv 0 \pmod{17}$

8. Find the remainder when:

   (a) $(9! \cdot 16 + 4311)^{8603}$ is divided by 11
   (b) $42! + 7^{28} + 66$ is divided by 29

9. If a is a natural number and p is a prime number, show that $a^p + a \cdot (p-1)!$ is divisible by p.
10. Find the remainder that $100 + 2^{33} + 16! + 29!$ leaves upon division by 19.

Challenging Problems

11. Show that a natural number $n > 1$ is prime if and only if n divides $(n - 2)! - 1$.
12. Show that, if p is a prime number and a and b are natural numbers, then

$$(a + b)^p \equiv a^p + b^p \pmod{p}$$

13. Prove that, for all primes $p > 2$, $(p - 2)! \equiv 1 \pmod{p}$.
14. Prove that, for all primes $p > 3$, $2 \cdot (p - 3)! \equiv -1 \pmod{p}$.
15. Is there a prime number p such that $(p - 1)! + 6$ is divisible by p?
16. Find all prime numbers p such that p divides $(p - 2)! + 6$.
17. Suppose $2^k + 1$ is a prime number. Prove that k has no prime divisors other than 2.
    [Hint: If $k = ab$ with b odd, consider $2^k + 1$ modulo $2^a + 1$.]
18. Prove that $a^{q-1} \equiv 1 \pmod{pq}$ if p and q are distinct primes such that $p - 1$ divides $q - 1$ and neither p nor q divides a.

Chapter 6
Sending and Receiving Secret Messages

Since ancient times, people have devised ways of sending secret messages to each other. Much of the original interest was for military purposes: commanders of one section of an army wanted to send messages to commanders of other sections of their army in such a way that the message could not be understood by enemy soldiers who might intercept it.

Some of the current interest in secret messages is still for military and similarly horrible purposes. However, there are also many other kinds of situations in which it is important to be able to send secret messages. For example, a huge amount of information is communicated via the internet. It is important that some of that information remain private, known only to the sender and recipient. One common situation is making withdrawals from bank accounts over the internet. If someone else was able to intercept the information being sent, that interceptor could transfer funds from the sender's bank account to the interceptor's bank account. There are many other commercial and personal communications that are sent electronically that people wish to keep secret.

"Cryptography" refers to techniques for reconfiguring messages so that they cannot be understood except by the intended recipient. *Encrypting* a message is the process of reconfiguring it; *decrypting* is the process of obtaining the original message from the encrypted one. For a method of cryptography to be useful, it must be the case that it would be virtually impossible (or at least extremely difficult) for anyone other than the intended recipient to be able to decrypt the messages.

A fundamental problem is that the intended recipient must have the information that is needed to decrypt encrypted messages. If the sender has to send the decrypting information to the recipient, unintended interceptors (e.g., someone who wants to transfer your money to his or her bank account) might get access to the method of decrypting as that method is being transmitted to the intended recipient. It can be very difficult to get the method of decrypting to the intended recipient while making sure that no one can intercept it on route.

© Springer Nature Switzerland AG 2018
D. Rosenthal et al., *A Readable Introduction to Real Mathematics*,
Undergraduate Texts in Mathematics, https://doi.org/10.1007/978-3-030-00632-7_6

The techniques of encrypting and decrypting messages for a given procedure are called the "keys" for that procedure. There must be a "key" for encrypting messages and a "key" for decrypting them.

Beginning in the 1940s, many people wondered whether there could be public key cryptography. That means, a method of doing cryptography that has the property that everyone in the world (the "public") can be told how to send the recipient an encrypted message. On the other hand, the recipient must be the only one who can decrypt messages sent using that procedure. That is, *public key cryptography* refers to methods of sending messages that allow the person who wishes to receive messages to publicly announce the way messages should be encrypted in such a way that only the person making the announcement can decrypt the messages. This seems to be impossible. If people know how to encrypt messages, won't they necessarily also be able to figure out how to decrypt them, just by reversing the encrypting procedure?

6.1   The RSA Method

Several methods for public key cryptography have been discovered. To actually use these methods requires computing with very large numbers. Thus, the methods would not be feasible without computers. One method is called "RSA" after three of the people who played important roles in its development: Ron Rivest, Adi Shamir, and Leonard Adleman. The only mathematics that is required to establish that the RSA method works is Fermat's Little Theorem (5.1.2).

Here is an outline of the method. The recipient announces to the entire world the following way to send messages. If you want to send a message, the first thing that you must do is to convert the message into a natural number. There are many ways of doing that; here is a rough description of one possibility. Write your message out as sentences in, say, the English language. Then convert the sentences into a natural number as follows. Let $A = 11$, $B = 12$, $C = 13, \ldots, Z = 36$. Let 37 represent a space. Let 38 represent a period, 39 a comma, 40 a semicolon, 41 a full colon, 42 an exclamation point, and 43 an apostrophe. If desired, other symbols could be represented by other two-digit natural numbers. Convert your English language message into a number by replacing each of the elements of your sentences by their corresponding numbers in the order that they appear. For any substantial message, this will result in a large natural number. Everyone would be able to reconstruct the English language message from that number if this procedure was known to them. For example, the sentence

PUBLIC KEY CRYPTOGRAPHY IS NEAT.

would be represented by the number

2631122219133721153537132835263025172811261835371929372415113038

Furthermore, if you read the rest of this chapter

$$352531432222372124253337331835371930373325282129 38$$

The RSA technique is a method for encrypting and decrypting numbers. Both the recipient and those who send messages must use computers to do the computations that are required; the numbers involved in any application of the technique that could realistically protect messages are much too large for the computations to be done by hand.

RSA encryption proceeds as follows. The person who wishes to receive messages, the recipient, chooses two very large prime numbers p and q that are different from each other, and then defines N to be pq. The recipient publicly announces the number N. However, the recipient keeps p and q secret. If p and q are large enough, it is not feasible for anyone other than the recipient to find p or q simply from knowing N; factoring very large numbers is an extremely difficult problem for even the most powerful current computers. There are some very large known prime numbers; such can easily be chosen so that the resulting $N = pq$ is impossible to factor in any reasonable amount of time. The recipient announces another natural number E, which we will call the *encryptor*, in addition to N. Below we will explain ways of choosing suitable E's.

The recipient then instructs all those who wish to send messages to do the following. Write your message as a natural number as described above. Let's say that M is the number representing your message. For this method to work, M must be less than N. If M is greater than or equal to N, you could divide your message into several smaller messages, each of which correspond to natural numbers less than N. The method we shall describe only works when M is less than N.

"To send me messages," the recipient announces to the world, "take your message M and compute the remainder that M^E (i.e., M raised to the power E) leaves upon division by N, and send me that remainder."

In other words, to send a message M, the sender computes the R between 0 and $N - 1$ such that $M^E \equiv R \pmod{N}$. The sender then sends R to the recipient.

How can the message be decrypted? That is, how can the recipient recover the original message M from R? This will require finding a *decryptor*, which will be possible for anyone who knows the factorization of N as the product pq but virtually impossible for anyone else. We shall see that, if E is chosen properly, there is a decryptor D such that for every integer L between 0 and $N-1$, $L^{ED} \equiv L \pmod{N}$. For such a D, since $R \equiv M^E \pmod{N}$, it follows that $R^D \equiv M^{ED} \pmod{N}$, and therefore, since $M^{ED} \equiv M \pmod{N}$, $R^D \equiv M \pmod{N}$. Thus, the recipient decrypts the message by finding the remainder that R^D leaves upon division by N.

Before discussing how to find encryptors E and decryptors D and why this method works, let's look at a simple example. In this example, the numbers are so small that anyone could figure out what p and q are, so this example could not realistically be used to encrypt messages. However, it illustrates the method.

Example 6.1.1. Let $p = 7$ and $q = 11$ be the primes; then $N = pq = 77$. Suppose that $E = 13$; as we shall see, there are always many possible values for E. Below we will discuss the properties that E must have. There is a technique for finding D, based on knowing p and q, that we shall describe later; that technique will produce $D = 37$ in this particular example.

In this example, the recipient announces $N = 77$ and $E = 13$ to the general public; the recipient keeps the values of p, q, and D secret.

The recipient instructs the world how to send messages. Suppose you want to send the message $M = 71$. Following the encryption rule, you must compute the remainder that $M^E = 71^{13}$ leaves upon division by 77. This is equivalent to calculating 71^{13} (mod 77). Let's compute that as follows, using some of the facts about modular arithmetic that we learned in the previous chapters. First, $71 \equiv -6$ (mod 77), so $M^E \equiv (-6)^{13}$ (mod 77). Now $6^3 = 216$ and $216 \equiv -15$ (mod 77), so $6^6 \equiv (6^3)^2 \equiv (-15)^2 \equiv 225 \equiv -6$ (mod 77). Therefore,

$$(-6)^{13} \equiv -6 \cdot (-6)^{12} \ (\text{mod } 77)$$
$$\equiv -6 \cdot 6^{12} \ (\text{mod } 77)$$
$$\equiv -6 \cdot (6^6)^2 \ (\text{mod } 77)$$
$$\equiv -6 \cdot (-6)^2 \ (\text{mod } 77)$$
$$\equiv -6^3 \ (\text{mod } 77)$$
$$\equiv 15 \ (\text{mod } 77)$$

Thus, the encrypted version of your message is 15.

Anyone who sees that the encrypted version is 15 would be able to discover your original message if they knew the decryptor. But the recipient is the only one who knows the decryptor.

In this special, easy, example, the recipient receives 15 and proceeds to decrypt it, using the decryptor 37, as follows. Your original message will be the remainder that 15^{37} leaves upon division by 77. Compute: $15^2 \equiv -6$ (mod 77). Then, $15^{26} \equiv (-6)^{13}$ (mod 77), which (as we saw above) is congruent to 15 (mod 77). Also, from $15^2 \equiv -6$ (mod 77) we obtain $15^8 \equiv (-6)^4 \equiv 6 \cdot 6^3 \equiv 6 \cdot (-15) \equiv -90 \equiv 64$ (mod 77). Therefore,

$$15^{37} \equiv 15^{26} \cdot 15^8 \cdot 15^3 \ (\text{mod } 77)$$
$$\equiv 15 \cdot 64 \cdot 15^3 \ (\text{mod } 77)$$
$$\equiv 64 \cdot 15^4 \ (\text{mod } 77)$$
$$\equiv 64 \cdot (15^2)^2 \ (\text{mod } 77)$$
$$\equiv 64 \cdot (-6)^2 \ (\text{mod } 77)$$
$$\equiv (-13) \cdot 36 \ (\text{mod } 77)$$
$$\equiv -468 \ (\text{mod } 77)$$

Of course, -468 is congruent to -468 plus any multiple of 77. Now $7 \cdot (77) = 539$. Hence, $15^{37} \equiv -468 \equiv -468 + 539 \equiv 71$ (mod 77). Therefore, we have decrypted the received message, 15, and obtained the original message, 71. (The number 71 must be the original message, since it is the only natural number less than N that is congruent to 71 modulo N.)

The above looks somewhat complicated. We now proceed to explain and analyze the method in more detail.

For p and q distinct primes and $N = pq$, we use the notation $\phi(N)$ to denote $(p-1)(q-1)$. (This is a particular case of a more general concept, known as the *Euler ϕ function*, that we will introduce in the next chapter.) The theorem that underlies the RSA technique is an easy consequence of Fermat's Little Theorem (5.1.2).

Theorem 6.1.2. *Let $N = pq$, where p and q are distinct prime numbers, and let $\phi(N) = (p-1)(q-1)$. If k and a are any natural numbers, then $a \cdot a^{k\phi(N)} \equiv a$ (mod N).*

Proof. The conclusion of the theorem is equivalent to the assertion that N divides the product of a and $a^{k(p-1)(q-1)} - 1$. Since N is the product of the distinct primes p and q, this is equivalent to the product being divisible by both p and q, for a natural number is divisible by both p and q if and only if its canonical factorization (Corollary 4.1.2) includes both of the primes p and q.

Consider p (obviously the same proof works for q). There are two cases. First, if p divides a, then p certainly divides $a \cdot (a^{k(p-1)(q-1)} - 1)$. If p does not divide a, then, by Fermat's Little Theorem (5.1.2), $a^{p-1} \equiv 1$ (mod p). Raising both sides of this congruence to the power $k(q-1)$ shows that $a^{k(p-1)(q-1)} \equiv 1$ (mod p). Thus, p divides $a^{k(p-1)(q-1)} - 1$, so it also divides $a \cdot (a^{k(p-1)(q-1)} - 1)$. This establishes the result in the case that p does not divide a. Thus, in both cases, p divides $(a \cdot a^{k(p-1)(q-1)} - a)$. Therefore, $a \cdot a^{k\phi(N)} \equiv a$ (mod N). $\qquad\square$

How does this theorem apply to the RSA method? We pick as an encryptor, E, any natural number that does not have any factor in common with $\phi(N)$. As we shall see in the next chapter, this implies that there is a natural number D such that ED is equal to the sum of 1 and a multiple of $\phi(N)$; that is, there is a D such that $ED = 1 + k\phi(N)$ for some natural number k. The theorem we have just proven shows that D is a decryptor, as follows. Suppose that M is the original message, so that $R \equiv M^E$ (mod N) is its encryption. Since R is congruent to M^E modulo N, R^D is congruent to M^{ED} modulo N. But $ED = 1 + k\phi(N)$, so R^D is congruent to $M^{1+k\phi(N)}$ modulo N. By the above theorem, $M^{1+k\phi(N)}$ is congruent to M modulo N. (Of course, M is a natural number less than N, which uniquely determines it.)

The explanation of how to find decryptors requires some additional mathematical tools that we develop in the next chapter. If N is very small, decryptors can be found simply by trial and error.

A complete description of the RSA technique, including choosing encryptors and finding decryptors, is given in the next chapter (see The RSA Procedure for Encrypting Messages 7.2.5).

6.2   Problems

Basic Exercises

1. You are to receive a message using the RSA system. You choose $p = 5$, $q = 7$, and $E = 5$. Verify that $D = 5$ is a decryptor. The encrypted message you receive is 17. What is the actual (decrypted) message?

2. Use the RSA system with $N = 21$ and the encryptor $E = 5$.

   (a) Encrypt the message $M = 7$.
   (b) Verify that $D = 5$ is a decryptor.
   (c) Decrypt the encrypted form of the message.

3. A person tries to receive messages without you being able to decrypt them. The person announces $N = 15$ and $E = 7$ to the world; the person uses such low numbers assuming that you don't understand RSA. An encrypted message $R = 8$ is sent. By trial and error, find a decryptor, D, and use it to find the original message.

Chapter 7
The Euclidean Algorithm and Applications

Each pair of natural numbers has a *greatest common divisor*; i.e., a largest natural number that is a factor of both of the numbers in the pair. For example, the greatest common divisor of 27 and 15 is 3, the greatest common divisor of 36 and 48 is 12, the greatest common divisor of 257 and 101 is 1, the greatest common divisor of 4 and 20 is 4 and the greatest common divisor of 7 and 7 is 7.

Notation 7.0.1. The *greatest common divisor* of the natural numbers a and b is denoted $\gcd(a, b)$.

Thus, $\gcd(27, 15) = 3$, $\gcd(36, 48) = 12$ and $\gcd(7, 21) = 7$.

One way to find the greatest common divisor of a pair of natural numbers is by factoring the numbers into primes. Then the greatest common divisor of the two numbers is obtained in the following way: For each prime that occurs as a factor of both numbers, find the highest power of that prime that is a common factor of both numbers and then multiply all those primes to all those powers together to get the greatest common divisor. For example, since $48 = 2^4 \cdot 3$ and $56 = 2^3 \cdot 7$, $\gcd(24, 56) = 2^3 = 8$. As another example, note that $\gcd(1292, 14440) = 76$, since $1292 = 2^2 \cdot 17 \cdot 19$ and $14440 = 2^3 \cdot 5 \cdot 19^2$ and $2^2 \cdot 19 = 76$.

Another way of finding the greatest common divisor of two natural numbers is by using what is called the *Euclidean Algorithm*. One advantage of this method is that it provides a way of expressing the greatest common divisor as a combination of the two original numbers in a way that can be extremely useful. In particular, this technique will allow us to compute a decryptor for each encryptor chosen for RSA encrypting. As we shall see, other applications of the Euclidean Algorithm include a method for finding integer solutions of linear equations in two variables (*Diophantine equations*) and a different proof of the Fundamental Theorem of Arithmetic.

© Springer Nature Switzerland AG 2018

D. Rosenthal et al., *A Readable Introduction to Real Mathematics*,
Undergraduate Texts in Mathematics, https://doi.org/10.1007/978-3-030-00632-7_7

7.1   The Euclidean Algorithm

The Euclidean Algorithm is based on the ordinary operation of division of natural numbers, allowing for a remainder. As we have seen, we can express that concept of division as follows: if a and b are any natural numbers, then there exist nonnegative integers q and r such that $a = bq + r$ and $0 \leq r < b$. (Recall that the number q is called the *quotient* and the number r is called the *remainder* in this equation.) If b divides a, then, of course, $r = 0$.

Let a and b be natural numbers. The Euclidean Algorithm for finding the greatest common divisor of a and b is the following technique. If $b = a$, then clearly the greatest common divisor is a. Suppose that b is less than a. (If b is greater than a, interchange the roles of a and b in what follows.) Divide a by b as described above to get q and r satisfying $a = bq + r$ with $0 \leq r < b$. If $r = 0$, then clearly the greatest common divisor of a and b is b itself. If r is not 0, divide r into b, to get $b = rq_1 + r_1$, where $0 \leq r_1 < r$. If $r_1 = 0$, stop here. If r_1 is different from 0, divide r_1 into r to get $r = r_1 q_2 + r_2$, where $0 \leq r_2 < r_1$. Continue this process until there is the remainder 0. (That will have to occur eventually since the remainders are all nonnegative integers and each one is less than the preceding one.) Thus, there is a sequence of equations as follows:

$$a = bq + r$$
$$b = rq_1 + r_1$$
$$r = r_1 q_2 + r_2$$
$$r_1 = r_2 q_3 + r_3$$
$$\vdots$$
$$r_{k-3} = r_{k-2} q_{k-1} + r_{k-1}$$
$$r_{k-2} = r_{k-1} q_k + r_k$$
$$r_{k-1} = r_k q_{k+1}$$

We show that r_k is the greatest common divisor of the original a and b. To see this, note first that r_k is a common divisor of a and b. This can be seen by "working your way up" the equations. Replacing r_{k-1} by $r_k q_{k+1}$ in the next to last equation gives $r_{k-2} = r_k q_{k+1} q_k + r_k = r_k(q_{k+1}q_k + 1)$. Thus, r_k divides r_{k-2}. The equation for r_{k-3} can then be rewritten:

$$r_{k-3} = r_k(q_{k+1}q_k + 1)q_{k-1} + r_k q_{k+1} = r_k\big((q_{k+1}q_k + 1)q_{k-1} + q_{k+1}\big)$$

Thus, r_{k-3} is also divisible by r_k. Continuing to work upwards eventually shows that r_k divides r, then b, and then a. Therefore, r_k is a common divisor of a and b.

To show that r_k is the greatest common divisor of a and b, we show that every other common divisor of a and b divides r_k. Suppose, then, that d is a natural number that divides both a and b. The equation $a = bq + r$ shows that d also divides r. Since

d divides both b and r, it divides r_1; since it divides r and r_1, it divides r_2; and so on. Eventually, we see that d also divides r_k. Hence, every common divisor of a and b divides r_k, so r_k is the greatest common divisor of a and b.

Let's look at an example. Suppose we want to use the Euclidean Algorithm to find the greatest common divisor of 33 and 24. We begin with $33 = 24 \cdot 1 + 9$. Then, $24 = 9 \cdot 2 + 6$. Then, $9 = 6 \cdot 1 + 3$. Then, $6 = 3 \cdot 2$. Thus, the greatest common divisor of 33 and 24 is 3.

Definition 7.1.1. A *linear combination* of the integers a and b is an expression of the form $ax + by$, where x and y are integers. We say that the integer d is a *linear combination of the integers a and b* if there exist integers x and y such that $ax + by = d$.

Obtaining the greatest common divisor by the Euclidean Algorithm allows us to express the greatest common divisor as a linear combination of the original numbers, as follows. First consider the above example. From the next to last equation, we get $3 = 9 - 6 \cdot 1$. Substituting the expression for 6 obtained from the previous equation into this one gives

$$3 = 9 - (24 - 9 \cdot 2) \cdot 1 = 9 - 24 + 9 \cdot 2 = 9 \cdot 3 - 24$$

Then solve for 9 in the first equation, getting $9 = 33 - 24 \cdot 1$, and substitute this into the above equation to get $3 = (33 - 24 \cdot 1) \cdot 3 - 24 = 33 \cdot 3 - 24 \cdot 4$. Therefore, $3 = 33 \cdot 3 + 24(-4)$. The greatest common divisor of the numbers 33 and 24, 3, is expressed in the last equation as a linear combination of 33 and 24.

The Euclidean Algorithm can always be used, as in the above example, to write the greatest common divisor of two natural numbers as a linear combination of those numbers. That is, given natural numbers a and b with greatest common divisor d, there exist integers x and y such that $d = ax + by$. This can be seen by working upwards in the sequence of equations that constitute the Euclidean Algorithm, as in the above example. The next to last equation can be used to write the greatest common divisor, r_k, as a linear combination of r_{k-1} and r_{k-2}; simply solve the next to last equation for r_k. Solving for r_{k-1} in the previous equation and substituting represents r_k as a linear combination of r_{k-2} and r_{k-3}. By continuing to work our way up the equations in the Euclidean Algorithm, we eventually obtain r_k as a linear combination of the given numbers a and b.

7.2 Applications

Definition 7.2.1. The integers a and b are said to be *relatively prime* if their only common divisor is 1; that is, if $\gcd(a, b) = 1$.

By the above-described consequence of the Euclidean Algorithm, $\gcd(a, b) = 1$ implies that there exist integers x and y such that $ax + by = 1$. This fact forms

the basis for a different proof of the Fundamental Theorem of Arithmetic (4.1.1).
We begin by using this fact to prove the following lemma. (This is a restatement of
Corollary 4.1.3; however, it is presented with a new and independent proof.)

Lemma 7.2.2. *If a prime number divides the product of two natural numbers, then
it divides at least one of the numbers.*

Proof. Suppose that p is prime and p divides ab. If p divides a, then we are done.
So suppose that p does not divide a; we show that in this case p divides b. Since p
is prime, the only possible factors that a could have in common with p are 1 and p.
Therefore, a and p are relatively prime and so there exist integers x and y such that
$ax + py = 1$. Multiply through by b, getting $bax + bpy = b$. Since p divides ab,
it divides the left-hand side of this equation, so it must divide b.   □

We need a slightly stronger lemma which follows easily from the above.

Lemma 7.2.3. *For any natural number n, if a prime divides the product of n natural
numbers, then it divides at least one of the numbers.*

Proof. This is a simple consequence of the previous lemma and mathematical
induction. The previous lemma is the case $n = 2$. Suppose that the result is true
for n factors, where n is greater than or equal to 2. Assume that p is prime and
that p divides $a_1 a_2 \cdots a_{n+1}$. If p does not divide a_1, then by the case $n = 2$, p
divides $a_2 \cdots a_{n+1}$. Hence, by the inductive hypothesis, p divides at least one of
$a_2, a_3, \ldots, a_{n+1}$.   □

We are now able to present another proof of the Fundamental Theorem of
Arithmetic.

Theorem 7.2.4 (The Fundamental Theorem of Arithmetic). *The factorization of
a natural number greater than 1 into primes is unique except for the order of the
primes.*

Proof. If there were natural numbers with two distinct factorizations, then, by the
Well-Ordering Principle (2.1.2), there would exist a smallest such natural number,
say N. Then $N = p_1^{\alpha_1} p_2^{\alpha_2} \cdots p_k^{\alpha_k} = q_1^{\beta_1} q_2^{\beta_2} \cdots q_l^{\beta_l}$. Notice that N cannot be prime,
since there is only one way to express a prime. Since p_1 divides N, it divides
$q_1^{\beta_1} q_2^{\beta_2} \cdots q_l^{\beta_l}$. By Lemma 7.2.3, p_1 divides some q_j. Since p_1 and q_j are prime,
$p_1 = q_j$. Dividing both expressions for N by this common factor would then
yield a smaller natural number with two distinct factorizations. This contradiction
establishes the result.   □

The Euclidean Algorithm is used to find decryptors in the RSA cryptography system. Before explaining this in general, we illustrate it in the case of Example 6.1.1
from the previous chapter. In that example, we started with $p = 7$ and $q = 11$, so
that $N = 77$ and $\phi(N) = 6 \cdot 10 = 60$. We took the encryptor $E = 13$. The crucial
property of the encryptor is that it is relatively prime to $\phi(N)$. That is true in this
case; clearly the only common factor of 13 and 60 is 1. Since $\gcd(13, 60) = 1$, the
consequence of the Euclidean Algorithm discussed above implies that there exist

integers x and y such that $1 = 13x + 60y$, or $13x = 1 - 60y$. Note that if x and y satisfy this equation, then, for every m, $13(x + 60m) = 1 - 60(y - 13m)$, since this latter equation is obtained from the previous one by simply adding $13 \cdot 60m$ to both sides of the equation. Thus, if the original x was negative, we could choose a positive m large enough so that $x + 60m$ is positive. Therefore, without loss of generality, we can assume that x is positive, which forces y to be negative in the equation $13x = 1 - 60y$. Replace $-y$ by u; then $13x = 1 + 60u$, with x and u both positive integers. We will find such x and u using the Euclidean Algorithm. First, however, note that any such x is a decryptor. To see this, first note that, as in Example 6.1.1, M^{13} is congruent to the encrypted version of the message M. Thus, the encrypted version of the message to the power x is congruent to $(M^{13})^x = M^{13x} = M^{1+60u} = M \cdot M^{60u}$, which is congruent modulo 77 to M by Theorem 6.1.2.

To obtain a decryptor for this example, we begin by using the Euclidean Algorithm to find $\gcd(13, 60)$:

$$60 = 13 \cdot 4 + 8$$
$$13 = 8 \cdot 1 + 5$$
$$8 = 5 \cdot 1 + 3$$
$$5 = 3 \cdot 1 + 2$$
$$3 = 2 \cdot 1 + 1$$
$$2 = 1 \cdot 2$$

Thus, the greatest common divisor of 13 and 60 is 1. Of course, we knew that already; we chose 13 to be relatively prime to 60. The point of using the Euclidean Algorithm is that it allows us to express 1 as a linear combination of 13 and 60, as follows. From the above equation $3 = 2 \cdot 1 + 1$ we get $1 = 3 - 2$. Since $5 = 3 \cdot 1 + 2$, we have $1 = 3 - 2 = 3 - (5 - 3) = 3 \cdot 2 - 5$. Continuing by working our way up and collecting coefficients gives the following:

$$1 = 3 \cdot 2 - 5$$
$$= (8 - 5) \cdot 2 - 5$$
$$= 8 \cdot 2 - 5 \cdot 3$$
$$= 8 \cdot 2 - (13 - 8) \cdot 3$$
$$= 8 \cdot 5 - 13 \cdot 3$$
$$= (60 - 13 \cdot 4) \cdot 5 - 13 \cdot 3$$
$$= 60 \cdot 5 - 13 \cdot 23$$

Equivalently, $1 - 60 \cdot 5 = -13 \cdot 23$. We are not done. We must find positive integers k and D such that $1 + 60k = 13D$. For any integer m, adding $-13 \cdot 60m$ to both sides of the above equation gives $1 - 60 \cdot (5 + 13m) = 13 \cdot (-23 - 60m)$. Taking $m = -1$ in this equation gives $1 + 60 \cdot 8 = 13 \cdot 37$. Thus, 37 is a decryptor.

We have illustrated and proven the RSA technique. The following is a statement of what we have established.

The RSA Procedure for Encrypting Messages 7.2.5. *The recipient chooses (very large) distinct prime numbers p and q and lets N = pq and $\phi(N)$ = (p − 1)(q − 1). The recipient then chooses a natural number E (which we are calling the "encryptor" and is often called the "public exponent") greater than 1 that is relatively prime to $\phi(N)$. The pair of numbers (N, E) is called the "public key." The recipient announces the public key and states that any message M consisting of a natural number less than N can be sent as follows: Compute the natural number R less than N such that $M^E \equiv R$ (mod N). The encrypted message that is sent is the natural number R. The recipient decrypts the message by using the Euclidean Algorithm to find natural numbers D (which we are calling the "decryptor" and is often called the "private exponent") and k such that $1 + k\phi(N) = ED$. The pair of numbers (N, D) is called the "private key"; the recipient keeps D secret. The recipient then recovers the original message M as the natural number less than N that is congruent to M^{ED} (mod N).*

The technique that we used to find decryptors can be used to solve many other practical problems.

Definition 7.2.6. A *linear Diophantine equation* is an equation of the form $ax + by = c$, where a, b, and c are integers, and for which we seek solutions (x, y) with x and y integers.

Example 7.2.7. A store sells two different kinds of boxes of candies. One kind sells for 9 dollars a box and the other kind for 16 dollars a box. At the end of the day, the store has received 143 dollars from the sale of boxes of candy. How many boxes did the store sell at each price?

How can we approach this problem? If x is the number of the less expensive boxes sold and y is the number of the more expensive boxes sold, then the information we are given is

$$9x + 16y = 143$$

There are obviously an infinite number of pairs (x, y) of real numbers that satisfy this equation; the graph in the plane of the set of solutions is a straight line. However, we know more about x and y than simply that they satisfy that equation. We know that they are both nonnegative integers. Are there nonnegative integral solutions? Are there any integral solutions at all? Since 9 and 16 are relatively prime, the Euclidean Algorithm tells us that there exist integers s and t (possibly negative) satisfying $9s + 16t = 1$. Multiplying through by 143 gives $9(143s) + 16(143t) = 143$. Therefore, there are integral solutions. However, it is not immediately clear whether there are nonnegative integral solutions, which the actual problem requires. Let's investigate.

We will use the Euclidean Algorithm to find integral solutions to the equation $9s + 16t = 1$. We first use the Euclidean Algorithm to find the greatest common divisor (even though we know it already):

$$16 = 9 \cdot 1 + 7$$
$$9 = 7 \cdot 1 + 2$$
$$7 = 2 \cdot 3 + 1$$
$$2 = 1 \cdot 2$$

Working our way back upwards to express 1 as a linear combination of 9 and 16 gives

$$1 = 7 - 2 \cdot 3$$
$$= 7 - (9 - 7) \cdot 3$$
$$= 7 \cdot 4 - 9 \cdot 3$$
$$= (16 - 9) \cdot 4 - 9 \cdot 3$$
$$= 16 \cdot 4 - 9 \cdot 7$$

Therefore, $9(-7) + 16 \cdot 4 = 1$. Multiplying by 143 yields $9(-7 \cdot 143) + 16(4 \cdot 143) = 143$. Note that $7 \cdot 143 = 1001$ and $4 \cdot 143 = 572$. For any integer m, we can add and subtract $16 \cdot 9m$ to the left-hand side of the above equation; thus, for every integer m,

$$9(-1001 + 16m) + 16(572 - 9m) = 143$$

This gives infinitely many integer solutions; what about nonnegative solutions?

We require that $-1001 + 16m$ be at least 0. That is equivalent to $16m \geq 1001$, or $m \geq \frac{1001}{16}$. Thus, $m \geq 62.5625$. The smallest integer m satisfying this inequality is $m = 63$. When $m = 63$, $-1001 + 16m = 7$ and $572 - 9m = 5$. Thus, one pair of nonnegative solutions to the original equation is $x = 7$ and $y = 5$. Are there other nonnegative solutions? We will show that all the solutions of this equation are of the form $x = -1001 + 16m$ and $y = 572 - 9m$, for some integer m (see Example 7.2.11 below). To show that the only nonnegative solution is $(7, 5)$ we reason as follows. If we take the next possible m, $m = 64$, then the y we get is $572 - 9 \cdot 64 = -4$. Obviously, if m is even larger, $572 - 9m$ will be even more negative. Therefore, the only pair of nonnegative solutions to the original equation is $(7, 5)$. Thus, the store sold 7 of the cheaper boxes and 5 of the more expensive boxes of candy. □

The basic theorem about solutions of linear Diophantine equations is the following.

Theorem 7.2.8. *The Diophantine equation $ax + by = c$, with a, b, and c integers, has integral solutions if and only if $\gcd(a, b)$ divides c.*

Proof. Let $d = \gcd(a, b)$. If there is a pair of integers (x, y) satisfying the equation, then $ax + by = c$ and, since d divides both of a and b, it follows that d divides c. This proves the easy part of the theorem.

The converse is also easy, but only because of what we have learned about the Euclidean Algorithm. We used the Euclidean Algorithm to prove that there exists a pair (s, t) of integers satisfying $as + bt = d$. If d divides c, then there is a k satisfying $c = dk$. Let $x = sk$ and $y = tk$. Then clearly $ax + by = c$.   □

As we've seen in the example where we determined the number of boxes of each kind of candy sold (Example 7.2.7), it is sometimes important to be able to determine all the solutions of a Diophantine equation. We use the very easy fact that $(x + bm, y - am)$ is a solution of $ax + by = c$ whenever (x, y) is a solution. This follows since $a(x + bm) + b(y - am) = ax + abm + by - abm = ax + by$. This shows that a Diophantine equation has an infinite number of solutions if it has any solution at all. However, in some situations, such as the problem about determining the number of different kinds of boxes of candy that were sold, it is important to have a unique solution that satisfies some other condition of the problem (such as requiring that both of x and y be nonnegative). Theorem 7.2.10 below precisely describes all the solutions of a given linear Diophantine equation.

We require a lemma that generalizes the fact that if a prime divides a product, then it divides at least one of the factors (Lemma 7.2.2).

Lemma 7.2.9. *If s divides tu and s is relatively prime to u, then s divides t.*

Proof. The hypothesis implies that there exists an r such that $tu = rs$. Write the canonical factorization of u into primes, $u = p_1^{\alpha_1} p_2^{\alpha_2} \cdots p_k^{\alpha_k}$. Then,

$$t p_1^{\alpha_1} p_2^{\alpha_2} \cdots p_k^{\alpha_k} = rs$$

Imagine factoring both sides of this equation into a product of primes. By the Fundamental Theorem of Arithmetic (see 4.1.1 or 7.2.4), the factorization of the left-hand side into primes has to be the same as the factorization of the right-hand side. Since s is relatively prime to u, none of the primes comprising s are among the p_i. Thus, all the primes in s occur to at least the same power in the factorization of t, and therefore s divides t. This proves the lemma.   □

Theorem 7.2.10. *Let $\gcd(a, b) = d$. The linear Diophantine equation $ax + by = c$ has a solution if and only if d divides c. If d divides c and (x_0, y_0) is a solution, then the integral solutions of the equation are all the pairs $\left(x_0 + m \cdot \frac{b}{d}, y_0 - m \cdot \frac{a}{d}\right)$, where m assumes all integral values.*

Proof. We already established the first assertion, the criterion for the existence of a solution (Theorem 7.2.8). If (x_0, y_0) is a solution, it is easy to see that each of the other pairs is also a solution, for

$$a\left(x_0 + m \cdot \frac{b}{d}\right) + b\left(y_0 - m \cdot \frac{a}{d}\right) = ax_0 + m \cdot \frac{ab}{d} + by_0 - m \cdot \frac{ab}{d} = ax_0 + by_0 = c$$

All that remains to be proven is that there are no solutions other than those described in the theorem. To see this, suppose that (x_0, y_0) is a solution and that (x, y) is any other solution of $ax + by = c$. Since $ax_0 + by_0 = c$, we can subtract the second equation from the first to conclude that

$$a(x - x_0) + b(y - y_0) = 0$$

Bring one of the terms to the other side and divide both sides of this equation by d to get

$$\frac{a}{d}(x - x_0) = \frac{b}{d}(y_0 - y)$$

Note that $\frac{a}{d}$ and $\frac{b}{d}$ are relatively prime. (For if e was a common factor greater than 1, then $d \cdot e$ would be a common divisor of a and b greater than d.) Hence, by Lemma 7.2.9, $\frac{a}{d}$ divides $(y_0 - y)$ and $\frac{b}{d}$ divides $(x - x_0)$. That is, there are integers k and l such that $y_0 - y = k \cdot \frac{a}{d}$ and $x - x_0 = l \cdot \frac{b}{d}$. Equivalently, $y = y_0 - k \cdot \frac{a}{d}$ and $x = x_0 + l \cdot \frac{b}{d}$. For (x, y) to be a solution, we must have

$$a\left(x_0 + l \cdot \frac{b}{d}\right) + b\left(y_0 - k \cdot \frac{a}{d}\right) = c$$

Thus,

$$ax_0 + l \cdot \frac{ab}{d} + by_0 - k \cdot \frac{ba}{d} = c$$

Since $ax_0 + by_0 = c$, we get $l \cdot \frac{ab}{d} - k \cdot \frac{ba}{d} = 0$. Therefore, $l = k$. Call this common value m. Then,

$$x = x_0 + m \cdot \frac{b}{d}$$

$$y = y_0 - m \cdot \frac{a}{d}$$

This proves the theorem. □

Example 7.2.11. The uniqueness of the solution to the "candy boxes problem" (Example 7.2.7) follows from this theorem. In that example, $\gcd(9, 16) = 1$, so all the solutions are indeed of the form $(-1001 + 16m, 572 - 9m)$. □

There are many other interesting applications of the theorem concerning solutions of linear Diophantine equations (see, for example, the problems at the end of this chapter).

Recall that we used the notation $\phi(N)$ to denote $(p - 1)(q - 1)$ when we were describing the RSA technique with $N = pq$, where p and q were distinct prime numbers. This is a special case of notation for a useful general concept.

Definition 7.2.12. The *Euler ϕ function* is defined for natural numbers m by: $\phi(m)$ is equal to the number of integers in $\{1, 2, \ldots, m\}$ that are relatively prime to m.

Example 7.2.13. To compute $\phi(8)$, we consider the set $\{1, 2, 3, 4, 5, 6, 7, 8\}$. We get $\phi(8) = 4$, since 1, 3, 5, and 7 are the numbers in the set that are relatively prime to 8. Similarly, $\phi(7) = 6$, and $\phi(12) = 4$.   □

Theorem 7.2.14. *If p is prime, then $\phi(p) = p - 1$.*

Proof. Since p is prime, every number in $\{1, 2, \ldots, p - 1\}$ is relatively prime to p, so $\phi(p) = p - 1$.   □

In discussing the RSA technique, we used the notation $\phi(pq) = (p - 1)(q - 1)$ when p and q were distinct primes. This is consistent with the definition of ϕ we are now using.

Theorem 7.2.15. *If p and q are distinct primes, then $\phi(pq) = (p - 1)(q - 1)$.*

Proof. Suppose that p and q are primes with p less than q (since they are different, one of them is less than the other), and let $N = pq$. Clearly pq is not relatively prime to N. Thus, to find $\phi(N)$ we must determine how many numbers in the set $S = \{1, 2, 3, \ldots, p, \ldots, q, \ldots, pq - 1\}$ are relatively prime to N. If a number is not relatively prime to N, then it must be divisible by either p or q or both. However, an element k of S cannot be divisible by both p and q. For if it was, the canonical factorization (Corollary 4.1.2) of k would show that it was divisible by pq. This is impossible since every number in S is less than pq.

There are a total of $pq - 1$ numbers in S; how many multiples of p are there in S? The set S contains p, $2p$, $3p$, and so on, up to $(q - 1)p$, since qp is not in S. Thus, there are $q - 1$ multiples of p in S. Similarly, there are $p - 1$ multiples of q in S. Therefore, there is a total of $(q - 1) + (p - 1) = p + q - 2$ numbers in S that are not relatively prime to N. Since there are $pq - 1$ numbers in S, the number of numbers in S that are relatively prime to N is

$$pq - 1 - (p + q - 2) = pq - p - q + 1 = (p - 1)(q - 1)$$

Therefore, $\phi(N) = (p - 1)(q - 1)$.   □

There is a formula for $\phi(m)$ for any natural number m greater than 1, in terms of the canonical factorization of m into a product of primes (see Problem 27 at the end of this chapter).

Fermat's beautiful theorem that $a^{p-1} \equiv 1 \pmod{p}$ (5.1.2) (for primes p and natural numbers a that are not divisible by p) can be generalized to composite moduli. We require the following lemma that generalizes Theorem 5.1.1.

Lemma 7.2.16. *If a is relatively prime to m and ax $\equiv$ ay* (mod *m), then x $\equiv$ y* (mod *m*).

Proof. We are given that *m* divides *ax* − *ay*. That is, *m* divides *a*(*x* − *y*). By Lemma 7.2.9, *m* divides *x* − *y*. Thus, *x* $\equiv$ *y* (mod *m*). $\square$

Euler's Theorem 7.2.17. *If m is a natural number greater than 1 and a is a natural number that is relatively prime to m, then* $a^{\phi(m)} \equiv 1$ (mod *m*).

Proof. The proof is very similar to the proof of Fermat's Little Theorem (5.1.2). Let $S = \{r_1, r_2, \ldots, r_{\phi(m)}\}$ be the set of numbers in $\{1, 2, \ldots, m\}$ that are relatively prime to *m*. Then let $T = \{ar_1, ar_2, \ldots, ar_{\phi(m)}\}$. Clearly, no two of the numbers in *S* are congruent to each other, since no two of them have the same remainder when divided by *m*. Note also that no two of the numbers in *T* are congruent to each other, since $ar_i \equiv ar_j$ (mod *m*) would imply, by Lemma 7.2.16, that $r_i \equiv r_j$ (mod *m*). Moreover, each ar_i is relatively prime to *m* and therefore so is any number that ar_i is congruent to. Thus, the numbers in $\{ar_1, ar_2, \ldots, ar_{\phi(m)}\}$ are congruent, in some order, to the numbers in $\{r_1, r_2, \ldots, r_{\phi(m)}\}$. It follows, as in the proof of Fermat's Little Theorem, that the product of all the numbers in *T* is congruent to the product of all the numbers in *S*. That is,

$$a \cdot r_1 \cdot a \cdot r_2 \cdots a \cdot r_{\phi(m)} \equiv r_1 r_2 \cdots r_{\phi(m)} \ (\text{mod } m)$$

Since $r_1 r_2 \cdots r_{\phi(m)}$ is relatively prime to *m*, we can divide both sides of this congruence by that product (see Lemma 7.2.16) to get $a^{\phi(m)} \equiv 1$ (mod *m*). $\square$

Fermat's Little Theorem is a special case of Euler's.

Corollary 7.2.18 (Fermat's Little Theorem). *If p is a prime and p does not divide a, then* $a^{p-1} \equiv 1$ (mod *p*).

Proof. Since *p* is prime, the fact that *p* does not divide *a* means that *a* and *p* are relatively prime. Also, $\phi(p) = p - 1$. Thus, Fermat's Little Theorem follows from Euler's Theorem (7.2.17). $\square$

7.3 Problems

Basic Exercises

1. Find the greatest common divisor of each of the following pairs of integers in two different ways, by using the Euclidean Algorithm and by factoring both numbers into primes:

   (a) 252 and 198
   (b) 291 and 573
   (c) 1800 and 240
   (d) 52 and 135

2. For each of the pairs in Problem 1 above, write the greatest common divisor as a linear combination of the given numbers.

3. Find integers x and y such that $3x - 98y = 12$.

4. (a) Find a formula for all integer solutions of the linear Diophantine equation
   $$3x + 4y = 14.$$
   (b) Find all pairs of natural numbers that solve the above equation.

5. Let ϕ be Euler's ϕ function. Find:

   (a) $\phi(12)$ (e) $\phi(97)$
   (b) $\phi(26)$ (f) $\phi(73)$
   (c) $\phi(21)$ (g) $\phi(101 \cdot 37)$
   (d) $\phi(36)$ (h) $\phi(3^{100})$

6. Use the Euclidean Algorithm to find the decryptors in Problems 1, 2, and 3 in Chapter 6.

Interesting Problems

7. Use the Euclidean Algorithm (assisted by a calculator) to find the greatest common divisor of each of the following pairs of natural numbers:

   (a) 47,295 and 297
   (b) 77,777 and 2,891

8. Find the smallest natural number x such that $24x$ leaves a remainder of 2 upon division by 59.

9. A small theater has a student rate of $3 per ticket and a regular rate of $10 per ticket. Last night $243 was collected from the sale of tickets. There were more than 50 but less than 60 tickets sold. How many student tickets were sold?

10. A liquid comes in 17 liter and 13 liter cans. Someone needs exactly 287 liters of the liquid. How many cans of each size should the person buy?

11. Let a, b, and n be natural numbers. Prove that if a^n and b^n are relatively prime, then a and b are relatively prime.

12. Let a, b, m, and n be natural numbers with m and n greater than 1. Assume that m and n are relatively prime. Prove that if $a \equiv b \pmod{m}$ and $a \equiv b \pmod{n}$, then $a \equiv b \pmod{mn}$.

13. Let a and b be natural numbers.

   (a) Suppose there exist integers m and n such that $am + bn = 1$. Prove that a and b are relatively prime.
   (b) Prove that $5a+2$ and $7a+3$ are relatively prime for every natural number a.

14. Let p be a prime number. Prove that $\phi(p^2) = p^2 - p$.

15. The public key $N = 55$ and $E = 7$ is announced. The encrypted message 5 is received.

    (a) Find a decryptor, D, and prove that D is a decryptor.
    (b) Decrypt 5 to find the original message.

16. Find a multiplicative inverse of 2^{29} modulo 9.
17. Prove that a has a multiplicative inverse modulo m if and only if a and m are relatively prime.

Challenging Problems

18. Suppose that a and b are relatively prime natural numbers such that ab is a perfect square. Show that a and b are each perfect squares.
19. Show that if m and n are relatively prime and a and b are any integers, then there is an integer x that simultaneously satisfies the two congruences $x \equiv a$ (mod m) and $x \equiv b$ (mod n).
20. Generalize the previous problem as follows (this result is called the *Chinese Remainder Theorem*):
    If $\{m_1, m_2, \ldots, m_k\}$ is a collection of natural numbers greater than 1, each pair of which is relatively prime, and if $\{a_1, a_2, \ldots, a_k\}$ is any collection of integers, then there is an integer x that simultaneously satisfies all of the congruences $x \equiv a_j$ (mod m_j). Moreover, if x_1 and x_2 are both simultaneous solutions of all of those congruences, then $x_1 \equiv x_2$ (mod $m_1 m_2 \cdots m_k$).
21. Let p be an odd prime and let $m = 2p$. Prove that $a^{m-1} \equiv a$ (mod m) for all natural numbers a.
22. Let a and b be relatively prime natural numbers greater than or equal to 2. Prove that $a^{\phi(b)} + b^{\phi(a)} \equiv 1$ (mod ab).
23. Suppose that $a, b,$ and c are each natural numbers. Prove that there are at most a finite number of pairs of natural numbers (x, y) that satisfy $ax + by = c$.
24. Show that m is prime if there is an integer a such that $a^{m-1} \equiv 1$ (mod m) and $a^k \not\equiv 1$ (mod m) for every natural number $k < m - 1$.
25. Suppose that a and m are relatively prime and that k is the smallest natural number such that a^k is congruent to 1 modulo m. Prove that k divides $\phi(m)$.
26. For p a prime and k a natural number, show that $\phi(p^k) = p^k - p^{k-1}$.
27. (Very Challenging) If the canonical factorization of the natural number n into primes is

$$n = p_1^{k_1} \cdot p_2^{k_2} \cdots p_m^{k_m}$$

prove that

$$\phi(n) = \left(p_1^{k_1} - p_1^{k_1-1}\right) \cdot \left(p_2^{k_2} - p_2^{k_2-1}\right) \cdots \left(p_m^{k_m} - p_m^{k_m-1}\right)$$

Chapter 8
Rational Numbers and Irrational Numbers

The only numbers that we have discussed so far are the "whole numbers;" that is, the integers. There are many other interesting things that can be said about the integers, but, for now, we consider other numbers, the *rational numbers*, also known as "fractions," and then the *real numbers*.

8.1 Rational Numbers

Definition 8.1.1. A *rational number* is a number of the form $\frac{m}{n}$, where m and n are integers and $n \neq 0$.

Some examples of rational numbers are $\frac{3}{4}$, $\frac{-7}{23}$, $\frac{12}{-36}$, $\frac{1}{2}$, and $\frac{2}{4}$.

Wait a minute. Are $\frac{1}{2}$ and $\frac{2}{4}$ different rational numbers? They are not; they are two different expressions representing the same number. Similarly $\frac{12}{48} = \frac{1}{4}$, $\frac{-7}{3} = \frac{7}{-3}$, $\frac{16}{2} = \frac{8}{1}$, and so on. The condition under which two different expressions as quotients of integers represent the same rational number is the following.

Definition 8.1.2. The rational number $\frac{m_1}{n_1}$ is equal to the rational number $\frac{m_2}{n_2}$ when $m_1 n_2 = m_2 n_1$.

Thus, when we use the expression $\frac{1}{2}$, we recognize that we are representing a number that could also be denoted $\frac{2}{4}$, $\frac{-3}{-6}$, and so on.

Why don't we allow 0 denominators in the expressions for rational numbers? If we did allow 0 denominators, the arithmetic would be very peculiar. For example, $\frac{7}{0}$ would equal $\frac{-12}{0}$, since $7 \cdot 0 = -12 \cdot 0$. In fact, we would have $\frac{a}{0} = \frac{b}{0}$ for all integers a and b. It is not at all useful to have such peculiarities as part of our arithmetic, so we do not allow 0 to be a denominator of any rational number.

© Springer Nature Switzerland AG 2018
D. Rosenthal et al., *A Readable Introduction to Real Mathematics*,
Undergraduate Texts in Mathematics, https://doi.org/10.1007/978-3-030-00632-7_8

Definition 8.1.3. The expression $\frac{m}{n}$ for a rational number is said to be in *lowest terms* if m and n are relatively prime.

Note that a representation of a rational number in lowest terms can be obtained by starting with any representation of the rational number and "dividing out" all the common factors of the numerator and the denominator.

Notation 8.1.4. The set of all rational numbers is denoted by $\mathbb{Q}$.

The operations of multiplication and addition of rational numbers can be defined in terms of the operations on integers.

Definition 8.1.5. The product of the rational numbers $\frac{m_1}{n_1}$ and $\frac{m_2}{n_2}$, denoted $\frac{m_1}{n_1} \cdot \frac{m_2}{n_2}$ or simply $\frac{m_1}{n_1} \frac{m_2}{n_2}$, is the rational number

$$\frac{m_1 m_2}{n_1 n_2}$$

The sum of the rational numbers $\frac{m_1}{n_1}$ and $\frac{m_2}{n_2}$ is the rational number

$$\frac{m_1}{n_1} + \frac{m_2}{n_2} = \frac{m_1 n_2 + m_2 n_1}{n_1 n_2}$$

We can think of the integers as the rational numbers whose denominator is 1; we invariably write them without the denominator. For example, we write -17 for $\frac{-17}{1}$ (and also, of course, for $\frac{-34}{2}$, and so on). In particular, we write 0 for $\frac{0}{1}$ and 1 for $\frac{1}{1}$. Note that, from Definition 8.1.5, 0 and 1 are, respectively, additive and multiplicative identities for the rational numbers, as they are for the integers. That is, $\frac{m}{n} + 0 = \frac{m}{n}$ and $\frac{m}{n} \cdot 1 = \frac{m}{n}$, for every rational number $\frac{m}{n}$. Also note that, as is the case with the set of integers, every rational number has an additive inverse: $\frac{m}{n} + \frac{-m}{n} = \frac{0}{n} = 0$.

Definition 8.1.6. A *multiplicative inverse* for the number x is a number y such that $xy = 1$.

Of course, 0 has no multiplicative inverse, since 0 times any number is 0. If x and y are both integers and $xy = 1$, then x and y must both be 1 or -1. Hence, the only integers that have multiplicative inverses within the set of integers are the numbers 1 and -1. In the set of rational numbers, the situation is very different.

Theorem 8.1.7. *If $\frac{m}{n}$ is a rational number other than 0, then $\frac{m}{n}$ has a multiplicative inverse.*

Proof. If $\frac{m}{n} \neq 0$, then $m \neq 0$. Therefore, $\frac{n}{m}$ is also a rational number and $\frac{m}{n} \frac{n}{m} = \frac{mn}{nm} = \frac{1}{1} = 1$. Therefore, $\frac{n}{m}$ is a multiplicative inverse for $\frac{m}{n}$. □

8.2 Irrational Numbers

In a sense, all actual numerical computations, by human or electronic computers, are done with rational numbers. However, it is important, within mathematics itself and in using mathematics to understand the world, to have other numbers as well.

Example 8.2.1. Suppose that you walk one mile due east and then one mile due north. How far are you from your starting point? The straight line from your starting point to your final position is the hypotenuse of a right triangle (see Definition 11.3.2) whose legs are each one mile long. The length of the hypotenuse is the distance that you are from your starting point. If x denotes that distance, then the Pythagorean Theorem (see 11.3.6) tells us that $x^2 = 2$.

It is obviously useful to have *some* number that denotes that distance. Is there a rational number x such that $x^2 = 2$? This question can be rephrased: are there integers m and n with $n \neq 0$ such that $\left(\frac{m}{n}\right)^2 = 2$? This, of course, is equivalent to the question of whether there are integers m and n different from 0 that satisfy the equation $m^2 = 2n^2$. This is a very concrete question about integers; what is the answer?

Theorem 8.2.2. *There do not exist integers m and n with $n \neq 0$ such that $\left(\frac{m}{n}\right)^2 = 2$.*

Proof. If there exist any such m and n, then, of course, there would exist such m and n that are relatively prime. We will show that this assumption leads to a contradiction. From $\left(\frac{m}{n}\right)^2 = 2$, we get $m^2 = 2n^2$. The equation $m^2 = 2n^2$ implies that m^2 is an even number, since it is the product of 2 and another number. What about m itself? If m were odd, then $m - 1$ would have to be even, so $m - 1 = 2k$ for some integer k, or $m = 2k + 1$. It would follow from this that $m^2 = (2k + 1)^2 = 4k^2 + 4k + 1 = 2(2k^2 + 2k) + 1$, which is an odd number (since it is 1 more than a multiple of 2). Thus, if m were odd, m^2 would have to be odd. Since m^2 is even, we conclude that m is even.

We now proceed to prove that n is also even. We know that $m = 2s$ for some integer s, from which it follows that $m^2 = 4s^2$. Substituting $4s^2$ for m^2 in the equation $m^2 = 2n^2$ gives $4s^2 = 2n^2$, or $2s^2 = n^2$. Thus, n^2 is an even number and, reasoning as we did above for m, it follows that n itself is an even number.

Therefore, if $\left(\frac{m}{n}\right)^2 = 2$, then m and n are both divisible by 2. This contradicts the assumption that m and n are relatively prime. $\qquad\square$

We have proven that there is no rational number that satisfies the equation $x^2 = 2$. Is there any number that satisfies this equation? It would obviously be very important to have such a number, for the purpose of specifying the distance you would be from your starting point in Example 8.2.1 and for many other purposes. Mathematicians have developed what are called the *real numbers*; the real numbers include numbers for all possible distances. The real numbers can be put into correspondence with the points on a line by labeling one point "0" and marking points to the right of 0 with the distances that they are from 0 (using any fixed units). Points on the line to the left of 0 are labeled with corresponding negative real numbers. The resulting *real number line* looks like

$$-3 \quad -2 \quad -1 \quad 0 \quad 1 \quad 2 \quad 3$$

The set of real numbers and the arithmetical operations on them can be precisely constructed in terms of rational numbers. In fact, there are several ways to do that. None of the ways of constructing the real numbers in terms of the rational numbers are easy; they all require substantial development. There are two main approaches, one using *Cauchy sequences* and the other using *Dedekind cuts*. The Dedekind cuts approach is outlined in Problem 15 at the end of this chapter. For the present, we simply assume that the real numbers exist and that the arithmetical operations on them have the usual properties.

Notation 8.2.3. The set of all real numbers is denoted by $\mathbb{R}$.

It can be shown that there is a positive real number x such that $x^2 = 2$. This number is denoted $\sqrt{2}$ or $2^{\frac{1}{2}}$.

Definition 8.2.4. For y and n natural numbers, the nth *root of* y is defined to be the positive real number x such that $x^n = y$. This is denoted either $y^{\frac{1}{n}}$ or $\sqrt[n]{y}$. More generally, $y^{\frac{m}{n}}$ is defined to be $\left(y^{\frac{1}{n}}\right)^m$. It can be shown that, for each y, m and n, $y^{\frac{m}{n}}$ defines a unique real number.

Definition 8.2.5. A real number that is not a rational number is said to be *irrational*.

Theorem 8.2.2 shows that $\sqrt{2}$ is not a rational number and thus can be rephrased as follows.

Theorem 8.2.6. *The number $\sqrt{2}$ is irrational.*

The symbol $\sqrt{3}$ represents the positive real number satisfying $\left(\sqrt{3}\right)^2 = 3$; is $\sqrt{3}$ irrational too?

We can establish a more general result.

Theorem 8.2.7. *If p is a prime number, then $\sqrt{p}$ is irrational.*

Proof. The proof will be similar to that of the special case $p = 2$. Suppose that $\frac{m}{n}$ is a fraction, written in lowest terms, satisfying $\left(\frac{m}{n}\right)^2 = p$. Then $m^2 = pn^2$. Since $m^2 = pn^2$, p divides m^2. Thus, p divides the product $m \cdot m$, from which it follows that p divides m (see Corollary 4.1.3). Therefore, there is an integer s such that $m = ps$, which gives $(ps)^2 = pn^2$. Dividing both sides of this equation by p gives $ps^2 = n^2$. Thus, p divides the product $n \cdot n$ and we conclude that p divides n. But this contradicts the fact that $\frac{m}{n}$ is written in lowest terms. Therefore, $\sqrt{p}$ cannot be rational. $\square$

Of course, some natural numbers do have rational square roots. For example, $\sqrt{1} = 1$, $\sqrt{4} = 2$, and $\sqrt{289} = 17$. What about $\sqrt{6}$? More generally is there a natural number m such that $\sqrt{m}$ is rational but $\sqrt{m}$ is not an integer?

Theorem 8.2.8. *If the square root of a natural number is rational, then the square root is a natural number.*

Proof. Assume that N is a natural number and that the square root of N is rational; that is, we can write $\sqrt{N} = \frac{a}{b}$, where the fraction $\frac{a}{b}$ is written in lowest terms. Then $N = \left(\frac{a}{b}\right)^2$, and so $a^2 = Nb^2$. If p is a prime number that divides b, then p divides a^2. Hence, p divides a by Corollary 4.1.3. But this is impossible, since a and b have no common factors. Thus, b is a natural number that is not divisible by any prime number. In other words, $b = 1$. Therefore, $\sqrt{N} = a$, a natural number.   □

A natural number that is the square of a natural number is said to be a *perfect square*. The canonical factorizations of perfect squares have a distinctive form.

Theorem 8.2.9. *A natural number other than 1 is a perfect square if and only if every prime number in its canonical factorization occurs to an even power.*

Proof. Let n be a natural number. If the canonical factorization of n (see Corollary 4.1.2) is $n = p_1^{\alpha_1} p_2^{\alpha_2} \cdots p_k^{\alpha_k}$, then $n^2 = p_1^{2\alpha_1} p_2^{2\alpha_2} \cdots p_k^{2\alpha_k}$. The uniqueness of the factorization into primes implies that this expression is the canonical factorization of n^2. All the exponents are obviously even. This proves that the square of every natural number has the property that every exponent in its canonical factorization is even. The converse is even easier. For if $m = p_1^{2\alpha_1} p_2^{2\alpha_2} \cdots p_k^{2\alpha_k}$, then obviously $m = n^2$, where $n = p_1^{\alpha_1} p_2^{\alpha_2} \cdots p_k^{\alpha_k}$.   □

We can use canonical factorizations to study other roots as well.

Example 8.2.10. The number $\sqrt[3]{4}$ is irrational.

Proof. If $\sqrt[3]{4} = \frac{m}{n}$ with m and n integers, then $4n^3 = m^3$. Write this equation in terms of the canonical factorizations of m and n, getting

$$4\left(p_1^{\alpha_1} p_2^{\alpha_2} \cdots p_r^{\alpha_r}\right)^3 = \left(q_1^{\beta_1} q_2^{\beta_2} \cdots q_s^{\beta_s}\right)^3$$

So,

$$2^2 \cdot p_1^{3\alpha_1} p_2^{3\alpha_2} \cdots p_r^{3\alpha_r} = q_1^{3\beta_1} q_2^{3\beta_2} \cdots q_s^{3\beta_s}$$

The prime 2 must occur to a power that is a multiple of 3, since every prime on the right-hand side of this equation occurs to such a power. On the other hand, 2 occurs on the left-hand side of the equation to a power that is two more than a multiple of 3. The uniqueness of the factorization into primes implies that no such equation is possible.   □

Example 8.2.11. The number $\sqrt{3} + \sqrt{5}$ is irrational.

Proof. Suppose that $\sqrt{3} + \sqrt{5} = r$, with r a rational number. Then $\sqrt{3} = r - \sqrt{5}$. Squaring both sides of this equation gives

$$3 = \left(r - \sqrt{5}\right)^2 = r^2 - 2\sqrt{5}r + 5$$

From this it follows that $2\sqrt{5}r = r^2 + 2$ or $\sqrt{5} = \frac{r^2+2}{2r}$. But r rational implies that $\frac{r^2+2}{2r}$ is rational, which contradicts the fact that $\sqrt{5}$ is irrational (Theorem 8.2.7). □

There are many situations in which it is important to solve various kinds of equations. In particular, polynomial equations arise quite frequently.

Definition 8.2.12. A *polynomial with integer coefficients* is an expression of the form

$$a_n x^n + a_{n-1} x^{n-1} + \cdots + a_1 x + a_0$$

where n is a nonnegative integer and the a_i are integers with a_n different from 0 (we also include "constant polynomials"; that is, polynomials where $n = 0$ and a_0 is any integer). The number x_0 is a *root* (or *zero*) of a polynomial if the value of the polynomial obtained by replacing x by x_0 is 0.

Example 8.2.13. The polynomial $x^5 + x - 1$ has no rational roots.

Proof. Suppose that $\frac{m}{n}$ is a rational root, where $\frac{m}{n}$ is written in lowest terms. Since $\frac{m}{n}$ is a root, substituting $\frac{m}{n}$ into the polynomial yields $\left(\frac{m}{n}\right)^5 + \frac{m}{n} - 1 = 0$. Multiplying both sides by n^5 gives $m^5 + mn^4 - n^5 = 0$, or $m(m^4 + n^4) = n^5$. It follows that every prime divisor of m is a divisor of n^5 and, hence, also of n. Since m and n are relatively prime, this implies that m has no prime divisors. Thus, m is either 1 or -1. Similarly, the above equation yields $m^5 = n(n^4 - mn^3)$ from which it follows that every prime divisor of n divides m. Thus, n does not have any prime divisors, so n is either 1 or -1. Therefore, the only possible values of $\frac{m}{n}$ are 1 or -1. That is, the only possible rational roots of the polynomial are 1 and -1. However, substituting 1 and -1 for x does not yield 0. Therefore, neither 1 nor -1 is a root. Thus, the polynomial does not have any rational roots. □

There is a general theorem, whose proof is similar to the above example, that is often useful in determining whether or not polynomials have rational roots and may also be used to find such roots.

The Rational Roots Theorem 8.2.14. *If $\frac{m}{n}$ is a rational root of the polynomial $a_k x^k + a_{k-1} x^{k-1} + \cdots + a_1 x + a_0$, where the a_j are integers and m and n are relatively prime, then m divides a_0 and n divides a_k.*

Proof. Assuming that $\frac{m}{n}$ is a root gives

$$a_k \left(\frac{m}{n}\right)^k + a_{k-1} \left(\frac{m}{n}\right)^{k-1} + \cdots + a_1 \left(\frac{m}{n}\right) + a_0 = 0$$

Multiplying both sides of this equation by n^k produces the equation

$$a_k m^k + a_{k-1} m^{k-1} n + \cdots + a_1 m n^{k-1} + a_0 n^k = 0$$

It follows that

$$m \left(a_k m^{k-1} + a_{k-1} m^{k-2} n + \cdots + a_1 n^{k-1}\right) = -a_0 n^k$$

Since m and n are relatively prime, m and n^k are also relatively prime. On the other hand, m divides $-a_0 n^k$. Thus, by Lemma 7.2.9, m divides a_0. Similarly,

$$a_k m^k = - \left(a_{k-1} m^{k-1} n + \cdots + a_1 m n^{k-1} + a_0 n^k\right)$$

so,

$$a_k m^k = -n \left(a_{k-1} m^{k-1} + \cdots + a_1 m n^{k-2} + a_0 n^{k-1}\right)$$

Since n is relatively prime to m and n divides $a_k m^k$, it follows (by Lemma 7.2.9) that n divides a_k. This proves the theorem.   □

Example 8.2.15. Find all the rational roots of the polynomial $2x^3 - x^2 + x - 6$.

Proof. By the Rational Roots Theorem (8.2.14), every rational root $\frac{m}{n}$ in lowest terms has the property that n divides 2 and m divides 6. Thus, the only possible values of n are $1, -1, 2, -2$, and the only possible values of m are $6, -6, 3, -3, 2, -2, 1, -1$. The possible values of the quotient $\frac{m}{n}$ are therefore $6, -6, 3, -3, 2, -2, \frac{3}{2}, -\frac{3}{2}, 1, -1, \frac{1}{2}, -\frac{1}{2}$. We can determine which of these possible roots actually are roots by simply substituting them for x and seeing if the result is 0. In this example, the only rational root is $\frac{3}{2}$.   □

The following is a question with an interesting answer: Do there exist two irrational numbers such that one of them to the power of the other is rational? That is, can x^y be rational if x and y are both irrational? A natural case to consider is that of $\left(\sqrt{3}\right)^{\sqrt{2}}$. In fact, however, it is not at all easy to determine whether or not $\left(\sqrt{3}\right)^{\sqrt{2}}$ is rational. Nonetheless, this example can still be used to prove that the general question has an affirmative answer, as follows. Either $\left(\sqrt{3}\right)^{\sqrt{2}}$ is rational or it is irrational. If it is rational, it provides an example showing that the answer to the question is affirmative. If $\left(\sqrt{3}\right)^{\sqrt{2}}$ is irrational, let $x = \left(\sqrt{3}\right)^{\sqrt{2}}$ and $y = \sqrt{2}$. Then x^y is an irrational number to an irrational power. But

$$x^y = \left(\left(\sqrt{3}\right)^{\sqrt{2}}\right)^{\sqrt{2}} = \left(\sqrt{3}\right)^{\sqrt{2}\cdot\sqrt{2}} = \left(\sqrt{3}\right)^2 = 3$$

This gives an affirmative answer in this case as well. In other words, $(\sqrt{3})^{\sqrt{2}}$ answers our original question, whether it itself is rational or irrational. In fact, $(\sqrt{3})^{\sqrt{2}}$ is irrational, as follows from the Gelfond–Schneider Theorem, whose proof is way beyond the level of this book.

8.3 Problems

Basic Exercises

1. Use the Rational Roots Theorem (8.2.14) to find all rational roots of each of the following polynomials (some may not have any rational roots at all):

   (a) $x^2 + 5x + 2$
   (b) $2x^3 - 5x^2 + 14x - 35$
   (c) $x^{10} - x + 1$

2. Show that $\sqrt[3]{5}$ is irrational.
3. Show that $\sqrt{\frac{1}{2}}$ is irrational.
4. Is the sum of an irrational number and a rational number always irrational?
5. Is an irrational number to a rational power always irrational?
6. Is the sum of two irrational numbers always irrational?
7. Is $\sqrt[3]{\sqrt{49} + 1}$ irrational?
8. If y is irrational and x is any rational number other than 0, show that xy is irrational.

Interesting Problems

9. Determine whether each of the following numbers is rational or irrational and prove that your answer is correct:

   (a) $32^{\frac{2}{3}}$ (d) $\dfrac{\sqrt{7}}{\sqrt[3]{15}}$ (f) $\sqrt{\frac{3}{8}}$
   (b) $28^{\frac{2}{5}}$
                          (e) $\dfrac{\sqrt{63}}{\sqrt{28}}$ (g) $\sqrt[7]{\frac{8}{9}}$
   (c) $\dfrac{\sqrt{7}}{\sqrt{5}}$

10. Prove that $\sqrt[3]{3} + \sqrt{11}$ is irrational.

Challenging Problems

11. Prove that the following numbers are irrational:

    (a) $\sqrt{5} + \sqrt{7}$ (d) $\sqrt{3} + \sqrt{5} + \sqrt{7}$

    (b) $\sqrt[3]{4} + \sqrt{10}$ (e) $\sqrt{3} - \frac{\sqrt{5}}{17}$

    (c) $\sqrt[3]{5} + \sqrt{3}$

12. Suppose that a and b are odd natural numbers and $a^2 + b^2 = c^2$. Prove that c is irrational.

13. Let k be a natural number. Prove that if the kth root of a natural number is a rational number, then the kth root is a natural number.

14. Prove that if a and b are natural numbers and n is a natural number such that $n^{\frac{a}{b}}$ is rational, then $n^{\frac{a}{b}}$ is a natural number.

15. (Very challenging.) In this problem, we outline the *Dedekind cuts* approach to constructing the real numbers. In this approach, real numbers are defined as certain kinds of sets of rational numbers. The definition is the following. A *real number* is a nonempty proper subset of the set of rational numbers that does not have a greatest element and has the property that if a rational number t is in the set, then so are all rational numbers less than t. (A "proper subset" is a subset which is not the whole set.)

    (a) Each rational number must, of course, also be a real number; i.e., representable as such a set of rational numbers. If r is a rational number, the representation of r as a real number is as the set of all rational numbers that are less than r. Prove that such a representation is a real number according to the definition given above.

    (b) If S and T are real numbers as defined above, then $S + T$ is defined to be the set of all $s + t$ with s in S and t in T. Prove that $S + T$ is a real number (i.e., has the above properties).

    (c) Prove that addition of real numbers as defined above is commutative; that is, $S + T = T + S$ for all real numbers S and T.

    (d) Prove that addition of real numbers as defined above is associative; that is, $(S_1 + S_2) + S_3 = S_1 + (S_2 + S_3)$, for all real numbers S_1, S_2, and S_3.

    (e) If S is a real number, define $-S$ to be the set of all rational numbers t such that $-t$ is not in S and $-t$ is not the smallest rational number that is not in S. Prove that $-S$ is a real number whenever S is a real number.

    (f) Let $\mathcal{O}$ denote the real number corresponding to the rational number 0 (i.e., the set of all x in $\mathbb{Q}$ such that x is less than 0). Prove that $S + \mathcal{O} = S$, for every real number S.

    (g) Prove that $S + (-S) = \mathcal{O}$, for every real number S.

    (h) We say that the real number S is *positive* if S contains a rational number that is greater than 0. If S and T are positive real numbers, then the product ST is defined to be the union of the set of all rational numbers that are less than or equal to 0 together with the set of all rational numbers of the form

st, where s is a positive number in S and t is a positive number in $\mathcal{T}$. Prove that the product of two positive real numbers is a real number.

(i) If S is a real number, define $|S|$ to be S if S is a positive real number and $-S$ otherwise. Say that a real number is *negative* if it is not positive and is not $\mathcal{O}$. Prove that $|S|$ is positive for all S not equal to $\mathcal{O}$.

(j) If S and $\mathcal{T}$ are real numbers, define the product $S\mathcal{T}$ to be $-(|S||\mathcal{T}|)$ if one is negative and the other is positive, to be $|S||\mathcal{T}|$ if both are negative, and to be $\mathcal{O}$ if either is $\mathcal{O}$. Prove that multiplication of real numbers is commutative; that is, $S\mathcal{T} = \mathcal{T}S$, for all real numbers S and $\mathcal{T}$.

(k) Let $\mathcal{I}$ denote the real number corresponding to the rational number 1. Prove that the product of $\mathcal{I}$ and S is S, for every real number S.

(l) For a positive real number S, define $\frac{1}{S}$ to be the union of the set of rational numbers that are less than or equal to 0 and the set of rational numbers t such that $\frac{1}{t}$ is not in S and $\frac{1}{t}$ is not the smallest rational number not in S. Prove that $\frac{1}{S}$ is a real number whenever S is a positive real number.

(m) For S a negative real number, define $\frac{1}{S}$ to be $-\frac{1}{|S|}$. For S any real number other than $\mathcal{O}$, prove that the product of S and $\frac{1}{S}$ is $\mathcal{I}$.

(n) Prove that multiplication of real numbers as defined above is associative; that is, $(S_1 S_2)S_3 = S_1(S_2 S_3)$, for all real numbers S_1, S_2, and S_3.

(o) Prove that multiplication of real numbers is distributive over addition; that is, $S_1(S_2 + S_3) = S_1 S_2 + S_1 S_3$, for all real numbers S_1, S_2, and S_3.

(p) (Existence of the square root of 2.) Let $\mathcal{U}$ denote the union of the set of negative rational numbers and the set of all rational numbers x such that x^2 is less than 2. Prove that $\mathcal{U}$ is a real number and that the product $\mathcal{U}\mathcal{U}$ is the real number corresponding to the rational number 2.

A very nice and complete exposition of the Dedekind cuts construction of the real numbers can be found in "Calculus" by Michael Spivak (Publish or Perish, Inc., Houston, Texas), which also contains a beautiful treatment of the principles of calculus.

Chapter 9
The Complex Numbers

The set of real numbers is rich enough to be useful in a wide variety of situations. In particular, it provides a number for every distance. There are, however, some situations where additional numbers are required.

9.1 What is a Complex Number?

Let's consider the problem of finding roots for polynomial equations. Recall that polynomials are expressions such as $7x^2 + 5x - 3$, and $\sqrt{2}x^3 + \frac{5}{7}x$, and $x^7 - 1$. The general definition is the following.

Definition 9.1.1. A *polynomial* is an expression of the form

$$a_n x^n + a_{n-1} x^{n-1} + \cdots + a_1 x + a_0$$

where n is a natural number and the a_i are numbers with $a_n \neq 0$. We also allow *constant polynomials*; i.e., expressions that are just a single number a_0. The a_i are called the *coefficients* of the polynomial. The natural number n, the highest power to which x occurs in the polynomial, is called the *degree* of the polynomial. A constant polynomial is said to have degree 0.

Note that in the definition of polynomial we used x as the variable; this is very standard. However, it is often the case that other variables are used as well. For example, $z^3 - 4z + 3$ would be a polynomial in the variable z.

A polynomial defines a function; whenever a specific number is substituted for x, the resulting expression is a number. The values of x for which the polynomial is 0 have special significance.

© Springer Nature Switzerland AG 2018
D. Rosenthal et al., *A Readable Introduction to Real Mathematics*,
Undergraduate Texts in Mathematics, https://doi.org/10.1007/978-3-030-00632-7_9

Definition 9.1.2. A *root* or *zero* of the polynomial $a_n x^n + a_{n-1}x^{n-1} + \cdots + a_1 x + a_0$ is a number that when substituted for x makes

$$a_n x^n + a_{n-1}x^{n-1} + \cdots + a_1 x + a_0 = 0$$

For example, 2 is a root of the polynomial $x^2 - 4$, 3 is a root of the polynomial $5x^2 - 2x - 39$, $-\frac{7}{5}$ is a root of the polynomial $5x + 7$, and so on.

A very natural question is: Which polynomials have roots? All polynomials of degree 1 have roots: the polynomial $a_1 x + a_0$ has the root $-\frac{a_0}{a_1}$. What about polynomials of degree 2? A simple example is the polynomial $x^2 + 1$. No real number is a root of that polynomial, since x^2 is nonnegative for every real number x, and therefore $x^2 + 1$ is strictly greater than 0 for every real number x. If the polynomial $x^2 + 1$ is to have a root, it would have to be in a larger number system than that of the real numbers. Such a system was invented by mathematicians hundreds of years ago.

We use the symbol i to denote a root of the polynomial $x^2 + 1$. That is, we define i^2 to be equal to -1. We then combine this symbol i with real numbers, using standard manipulations of algebra in the usual ways, to get the "complex numbers." The definition is the following.

Definition 9.1.3. A *complex number* is an expression of the form $a + bi$ where a and b are real numbers. The real number a is called the *real part* of $a + bi$ and the real number b is called the *imaginary part* of $a + bi$. We sometimes use the notation $\text{Re}(z)$ and $\text{Im}(z)$ to denote the real and imaginary parts of the complex number z, respectively. Addition of complex numbers is defined by

$$(a + bi) + (c + di) = (a + c) + (b + d)i$$

Multiplication of complex numbers is defined by

$$(a + bi)(c + di) = ac + adi + bic + bdi^2$$
$$= ac + bdi^2 + (ad + bc)i$$
$$= (ac - bd) + (ad + bc)i$$

where we replaced i^2 by -1 to get the last equation.

Example 9.1.4.

$$(6 + 2i) + (-4 + 5i) = 2 + 7i$$

$$\left(-\sqrt{12} + \sqrt{6}i\right) + (4 + \pi i) = \left(-\sqrt{12} + 4\right) + \left(\sqrt{6} + \pi\right)i$$

$$(7+2i)(3-4i) = 21+6i-28i-8i^2 = 21-22i-8(-1) = 21+8-22i = 29-22i$$

Notation 9.1.5. The set of all complex numbers is denoted by $\mathbb{C}$.

We use the symbol 0 as an abbreviation for the complex number $0 + 0i$. More generally, we use a as an abbreviation for the complex number $a + 0i$. Thus, every real number is also a complex number. Similarly, we use bi as an abbreviation for the complex number $0 + bi$. When r is a real number, then $r(a + bi)$ is simply $ra + rbi$.

Note that every complex number has an additive inverse (i.e., a complex number that gives 0 when added to the given number). For example, the additive inverse of $-7 + \sqrt{2}i$ is $7 - \sqrt{2}i$. In general, the additive inverse of $a + bi$ is $-a + (-b)i$.

Definition 9.1.6. The number $a - bi$ is called the *complex conjugate* of the number $a + bi$. The complex conjugate of a complex number is often denoted by placing a horizontal bar over the complex number:

$$\overline{a + bi} = a - bi$$

Example 9.1.7. The complex conjugate of $2 + 3i$ is $2 - 3i$, or $\overline{2 + 3i} = 2 - 3i$. Similarly, $\overline{-\sqrt{3} - 5i} = -\sqrt{3} + 5i$, and $\overline{9} = 9$.

The product of a complex number and its conjugate is important.

Theorem 9.1.8. *For any complex number $a + bi$, $(a + bi)(a - bi) = a^2 + b^2$.*

Proof. Simply multiplying gives the result. $\qquad\square$

Definition 9.1.9. The *modulus* of the complex number $a + bi$ is $\sqrt{a^2 + b^2}$; it is often denoted $|a + bi|$.

Thus, $(a + bi)(\overline{a + bi}) = |a + bi|^2$.

Do complex numbers have multiplicative inverses? That is, given $a + bi$, is there a complex number $c + di$ such that $(a + bi)(c + di) = 1$? Of course, the complex number 0 cannot have a multiplicative inverse, since its product with any complex number is 0. What about other complex numbers?

Given a complex number $a + bi$, let's try to find a multiplicative inverse $c + di$ for it. Suppose that $(a + bi)(c + di) = 1$. Multiplying both sides of this equation by $\overline{a + bi}$ and using the fact that $(\overline{a + bi})(a + bi) = a^2 + b^2$ yields $(a^2 + b^2)(c + di) = a - bi$. Since $a^2 + b^2$ is a real number, this implies (unless $a^2 + b^2 = 0$) that $c + di = \frac{a}{a^2 + b^2} - \frac{b}{a^2 + b^2}i$. (Note that if $a^2 + b^2 = 0$, then $a = 0$ and $b = 0$, so the number $a + bi$ is 0.) Therefore, if $a + bi$ has a multiplicative inverse, that multiplicative inverse must be $\frac{a}{a^2 + b^2} - \frac{b}{a^2 + b^2}i$. In fact, as we now show, that expression is a multiplicative inverse for $a + bi$.

Theorem 9.1.10. *If $a + bi \neq 0$, then $\frac{a}{a^2 + b^2} - \frac{b}{a^2 + b^2}i$ is a multiplicative inverse for $a + bi$.*

Proof. We verify this by simply multiplying

$$(a + bi) \cdot \left(\frac{a}{a^2 + b^2} - \frac{b}{a^2 + b^2} i \right) = \frac{a^2}{a^2 + b^2} + \frac{b^2}{a^2 + b^2} - \frac{ab}{a^2 + b^2} i + \frac{ab}{a^2 + b^2} i$$

which simplifies to $\frac{a^2 + b^2}{a^2 + b^2} = 1$. □

As with real numbers, the multiplicative inverse of the complex number $a + bi$ is often denoted $\frac{1}{a+bi}$.

9.2   The Complex Plane

It is very useful to represent complex numbers in a coordinatized plane. We let the complex number $a + bi$ correspond to the point (a, b) in the ordinary xy-plane. Note that the modulus $|a + bi|$ is the distance from (a, b) to the origin. We will also use the angle that the line from the origin to (a, b) makes with the positive x-axis.

In day to day life, angles are usually measured in degrees: a right angle is $90°$, a straight angle is $180°$, and an angle of $37°$ is $\frac{37}{180}$ of a straight angle. For doing mathematics, however, it is often more convenient to measure angles differently.

Definition 9.2.1. The *radian measure* of the angle θ is the length of the arc of a circle of radius 1 that is cut off by an angle θ at the center of the circle. (See Figure 9.1.)

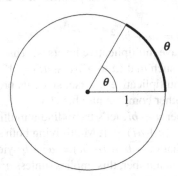

Fig. 9.1 The radian measure of an angle

Thus, since a circle of radius 1 has circumference 2π, the radian measure of a right angle is $\frac{\pi}{2}$, of a straight angle is π, of an angle of $60°$ is $\frac{\pi}{3}$, and so on. Note that 2π is a full revolution. We will use the radian measure of angles for the rest of this chapter.

Definition 9.2.2. For a complex number $a + bi$ other than 0, the *argument* of $a + bi$ is the angle from the positive x-axis in a counterclockwise direction to the line from $(0, 0)$ to (a, b). For any integer k, the angle $\theta + 2\pi k$ measured from the positive x-axis ends up at the same position as θ. Hence, if θ is an argument of a given complex number, and k is any integer, then $\theta + 2\pi k$ is also an argument of the complex number. We define the argument of 0 to be 0.

We require the basic properties of the trigonometric functions *sine* and *cosine*.

If the complex number $a + bi$ has modulus r and argument θ, then $a = r\cos\theta$, and $b = r\sin\theta$. To see this, first consider the case where both a and b are greater than 0, which is equivalent to saying that $0 < \theta < \frac{\pi}{2}$. Then the situation is as in Figure 9.2. The fact that the cosine of an angle in a right triangle is the length of its adjacent side divided by the length of its hypotenuse gives $\cos\theta = \frac{a}{r}$, or $a = r\cos\theta$. Similarly, the fact that the sine of θ is the length of the opposite side divided by the length of the hypotenuse gives $\sin\theta = \frac{b}{r}$, or $b = r\sin\theta$.

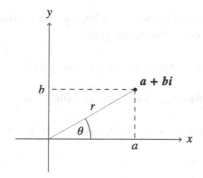

Fig. 9.2 Representation of a complex number with $a > 0$ and $b > 0$

The general case, for any real numbers a and b, requires the general definition of sine and cosine for all angles, not just for those that are less than a right angle. The circle with center at $(0, 0)$ and radius 1 is called the *unit circle*. If θ is an angle measured in a counterclockwise direction from the positive x-axis to the line from $(0, 0)$ to the point (x, y) on the unit circle, then $\cos\theta$ is defined to be x and $\sin\theta$ is defined to be y. For example, when θ is 0, the corresponding point on the unit circle is $(1, 0)$. Thus, $\cos(0) = 1$ and $\sin(0) = 0$. When θ equals $\frac{\pi}{2}$, the corresponding point on the unit circle is $(0, 1)$, so $\cos\left(\frac{\pi}{2}\right) = 0$ and $\sin\left(\frac{\pi}{2}\right) = 1$. One can similarly check that $\cos(\pi) = -1$, $\sin(\pi) = 0$, $\cos\left(\frac{3\pi}{2}\right) = 0$ and $\sin\left(\frac{3\pi}{2}\right) = -1$.

It is easy to see that, if a and b are both greater than 0, the generalized definitions of sine and cosine coincide with the original ones.

If $a + bi$ is a complex number other than 0, then its modulus, $r = \sqrt{a^2 + b^2}$, is not 0. The point $\left(\frac{a}{r}, \frac{b}{r}\right)$ lies on the unit circle. The line from $(0, 0)$ to (a, b) is the same line as the line from $(0, 0)$ to $\left(\frac{a}{r}, \frac{b}{r}\right)$ (since they both have the same slopes and they both pass through the origin). Let θ be the argument of $a+bi$. Since θ is also the angle from the positive x-axis to the line from $(0, 0)$ to $\left(\frac{a}{r}, \frac{b}{r}\right)$, the definitions yield $\cos\theta = \frac{a}{r}$ and $\sin\theta = \frac{b}{r}$. Thus, $r\cos\theta = a$ and $r\sin\theta = b$, from which it follows that $a + bi = r(\cos\theta + i\sin\theta)$. (The only complex number whose modulus is 0 is the number 0, and 0 is the only complex number whose argument is not defined.)

Definition 9.2.3. The *polar form* of the complex number with modulus r and argument θ is $r(\cos\theta + i\sin\theta)$.

One reason that the polar form is important is because there is a very nice description of multiplication of complex numbers in terms of their moduli and arguments.

Theorem 9.2.4. *The modulus of the product of two complex numbers is the product of their moduli. The argument of the product of two complex numbers is the sum of their arguments.*

Proof. Simply multiplying the two complex numbers $r_1(\cos\theta_1 + i\sin\theta_1)$ and $r_2(\cos\theta_2 + i\sin\theta_2)$ and collecting terms yields

$$r_1 r_2\big((\cos\theta_1\cos\theta_2 - \sin\theta_1\sin\theta_2) + i(\cos\theta_1\sin\theta_2 + \sin\theta_1\cos\theta_2)\big)$$

We require the addition formulae for cosine and sine:

$$\cos(\theta_1 + \theta_2) = \cos\theta_1\cos\theta_2 - \sin\theta_1\sin\theta_2$$

and

$$\sin(\theta_1 + \theta_2) = \sin\theta_1\cos\theta_2 + \sin\theta_2\cos\theta_1$$

Using these addition formulae in the above equation shows that the product is equal to

$$r_1 r_2\big(\cos(\theta_1 + \theta_2) + i\sin(\theta_1 + \theta_2)\big)$$

This proves the theorem.   □

Thus, to multiply two complex numbers, we can simply multiply their moduli and add their arguments. In particular, the case where the two complex numbers are equal to each other shows that the square of a complex number is obtained by squaring its modulus and doubling its argument. One application of this fact is the following.

Theorem 9.2.5. *Every complex number has a complex square root.*

Proof. To show that any given complex number has a square root, write the given number in polar form, say $z = r(\cos\theta + i\sin\theta)$. Let w equal $\sqrt{r}\left(\cos\frac{\theta}{2} + i\sin\frac{\theta}{2}\right)$. By the previous theorem (9.2.4), $w^2 = z$. □

It is also easy to compute powers higher than 2.

De Moivre's Theorem 9.2.6. *For every natural number n,*

$$\left(r(\cos\theta + i\sin\theta)\right)^n = r^n(\cos n\theta + i\sin n\theta)$$

Proof. This is easily established by induction on n. The case $n = 1$ is clear. Suppose that the formula holds for $n = k$; that is, suppose

$$\left(r(\cos\theta + i\sin\theta)\right)^k = r^k(\cos k\theta + i\sin k\theta)$$

Multiplying both sides of this equation by $r(\cos\theta + i\sin\theta)$ and using Theorem 9.2.4 gives

$$\left(r(\cos\theta + i\sin\theta)\right)^{k+1} = r^k(\cos k\theta + i\sin k\theta) \cdot r(\cos\theta + i\sin\theta)$$

$$= r \cdot r^k(\cos(k\theta + \theta) + i\sin(k\theta + \theta))$$

$$= r^{k+1}\left(\cos((k+1)\theta) + i\sin((k+1)\theta)\right)$$

This is the formula for $n = k + 1$, so the theorem is established by mathematical induction. □

De Moivre's Theorem leads to some very nice computations, such as the following.

Example 9.2.7. We can compute $(1 + i)^8$ as follows. First, $|1 + i| = \sqrt{2}$. Plotting $1+i$ as the point $(1, 1)$ in the plane makes it apparent that the argument of $1+i$ is $\frac{\pi}{4}$. Thus, by De Moivre's Theorem (9.2.6), the modulus of $(1+i)^8$ is $\left(\sqrt{2}\right)^8 = 2^4 = 16$ and the argument is $8 \cdot \frac{\pi}{4} = 2\pi$. It follows that

$$(1 + i)^8 = 16(\cos 2\pi + i\sin 2\pi) = 16$$

Therefore, $(1 + i)^8 = 16$.

The following is a very similar computation.

Example 9.2.8.

$$(1 + i)^{100} = \left(\sqrt{2}\left(\cos\frac{\pi}{4} + i\sin\frac{\pi}{4}\right)\right)^{100} = 2^{50}(\cos 25\pi + i\sin 25\pi)$$

Since the angle with the positive x-axis of 25π radians is in the same position as the angle of π radians, it follows that

$$(1+i)^{100} = 2^{50}(\cos \pi + i \sin \pi) = 2^{50}(-1+0) = -2^{50}$$

It is useful to have expressions for the cosines and sines of certain angles.

Example 9.2.9. Some particular values of cosine and sine are the following: $\cos \frac{\pi}{4} = \frac{\sqrt{2}}{2}$, $\sin \frac{\pi}{4} = \frac{\sqrt{2}}{2}$, $\cos \frac{\pi}{6} = \frac{\sqrt{3}}{2}$, $\sin \frac{\pi}{6} = \frac{1}{2}$, $\cos \frac{\pi}{3} = \frac{1}{2}$, and $\sin \frac{\pi}{3} = \frac{\sqrt{3}}{2}$.

Proof. To determine the trigonometric functions of $\frac{\pi}{4}$, simply note that a right triangle that has an angle of $\frac{\pi}{4}$ is isosceles. If such a right triangle has legs of length 1, the Pythagorean Theorem implies that the hypotenuse has length $\sqrt{2}$. Thus, $\cos \frac{\pi}{4}$ and $\sin \frac{\pi}{4}$ are both $\frac{1}{\sqrt{2}} = \frac{\sqrt{2}}{2}$.

For the remaining angles, begin with an equilateral triangle whose sides all have length 1. Then bisect one of the angles, as shown in Figure 9.3. This divides the equilateral triangle into two right triangles. Since the original triangle is equilateral, and the angles of a triangle sum to π, each angle of the original triangle is $\frac{\pi}{3}$. Thus, the smaller angles in each of the right triangles are each $\frac{\pi}{6}$. By the Pythagorean Theorem, the bisector of the angle has length $\frac{\sqrt{3}}{2}$. The values of cosine and sine for $\frac{\pi}{3}$ and $\frac{\pi}{6}$ are then immediate from the definitions, using either of the right triangles.   □

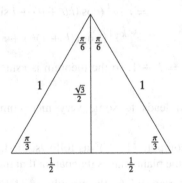

Fig. 9.3 Calculating the cosine and sine of $\frac{\pi}{3}$ and $\frac{\pi}{6}$

Using these values of cosine and sine, it is easy to calculate some related values. The following example is typical.

Example 9.2.10. We can compute $\cos \frac{2\pi}{3}$ and $\sin \frac{2\pi}{3}$ as follows. Place the angle so that one side is on the positive x-axis and the angle is measured counterclockwise from there. Let (x, y) be the corresponding point on the unit circle. Then $x = \cos \frac{2\pi}{3}$ and $y = \sin \frac{2\pi}{3}$. We use the right triangle pictured in Figure 9.4. Since $\cos \frac{\pi}{3} = \frac{1}{2}$, it follows that $x = -\frac{1}{2}$. Since $\sin \frac{\pi}{3} = \frac{\sqrt{3}}{2}$, it follows that $y = \frac{\sqrt{3}}{2}$.   □

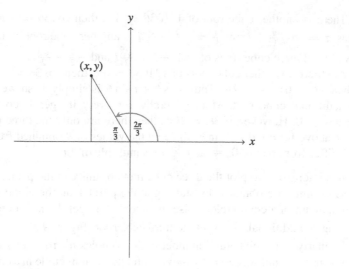

Fig. 9.4 Determining the cosine and sine of $\frac{2\pi}{3}$

It is interesting to compute the roots of the complex number 1. The number 1 is sometimes called *unity*.

Example 9.2.11 (Square Roots of Unity). Obviously, $1^2 = 1$ and $(-1)^2 = 1$. Are there any other complex square roots of 1?

To compute the square roots of 1 we can proceed as follows. Let $z = r(\cos\theta + i\sin\theta)$. By De Moivre's Theorem (9.2.6), $z^2 = r^2(\cos 2\theta + i\sin 2\theta)$. If $z^2 = 1$, then r^2 must be the modulus of 1; i.e., $r^2 = 1$. Since r is nonnegative, it follows that $r = 1$. Also, $\cos 2\theta + i\sin 2\theta = 1$. Therefore, $\cos 2\theta = 1$ and $\sin 2\theta = 0$. What are the possible values of θ? Clearly, $\theta = 0$ is one solution, as is $\theta = \pi$; the corresponding values of z are $z = \cos 0 + i\sin 0 = 1$ and $z = \cos\pi + i\sin\pi = -1$.

Are there any other possible values of θ? Of course there are: for example, θ could be 2π or 3π or 4π or 5π. If θ is any multiple of π, then $\cos 2\theta = 1$ and $\sin 2\theta = 0$. However, we do not get any new values of z by using those other values of θ. We only get $z = 1$ or $z = -1$ depending upon whether we have an even or an odd multiple of π. It is easily seen that only the multiples of π simultaneously satisfy the equations $\cos 2\theta = 1$ and $\sin 2\theta = 0$. (This follows from the fact that $\cos\phi = 1$ only when ϕ is a multiple of 2π, so $\cos 2\theta = 1$ only when θ is a multiple of π.) Thus, the only complex square roots of 1 are 1 and -1. □

Cube roots of unity are more interesting. The only real number z that satisfies $z^3 = 1$ is $z = 1$. However, there are other complex numbers satisfying this equation.

Example 9.2.12 (Cube Roots of Unity). Suppose that $z = r(\cos\theta + i\sin\theta)$ and $z^3 = 1$. Then clearly $r = 1$. By De Moivre's Theorem (9.2.6), $z^3 = \cos 3\theta + i\sin 3\theta$. From $z^3 = 1$ we get $\cos 3\theta = 1$ and $\sin 3\theta = 0$. These equations are, of course, satisfied by $\theta = 0$, which gives $z = \cos 0 + i\sin 0 = 1$, the obvious cube root of 1. But also $\cos 3\theta = 1$ and $\sin 3\theta = 0$ when $3\theta = 2\pi$. That is, when $\theta = \frac{2\pi}{3}$. Thus, $z = \cos\frac{2\pi}{3} + i\sin\frac{2\pi}{3} = -\frac{1}{2} + \frac{\sqrt{3}}{2}i$ is another cube root of 1.

There is another cube root of 1. If $3\theta = 4\pi$, then $\cos 3\theta = 1$ and $\sin 3\theta = 0$. Thus, $z = \cos \frac{4\pi}{3} + i \sin \frac{4\pi}{3} = -\frac{1}{2} - \frac{\sqrt{3}}{2}i$ is another cube root of 1. Therefore, we have found three cube roots of 1: $1, -\frac{1}{2} + \frac{\sqrt{3}}{2}i$ and $-\frac{1}{2} - \frac{\sqrt{3}}{2}i$.

Are there any other cube roots of 1? If $3\theta = 6\pi$, then $\cos 3\theta = 1$ and $\sin 3\theta = 0$. When $3\theta = 6\pi$, $\theta = 2\pi$. Thus, $\cos \theta + i \sin \theta$ is simply 1, so we are not getting an additional cube root. More generally, for every integer k, $\cos 2k\pi = 1$ and $\sin 2k\pi = 0$. However, if $3\theta = 2k\pi$, then there are only the three different values given above for $\cos \theta + i \sin \theta$, since all the values of θ obtained from other values of k differ from one of 0, $\frac{2\pi}{3}$ and $\frac{4\pi}{3}$ by a multiple of 2π.   □

It is interesting to plot the three cube roots of unity in the plane. The three cube roots of unity are obtained by starting at the point 1 on the circle of radius 1 and then moving in a counterclockwise direction $\frac{2\pi}{3}$ to get the next cube root and then moving an additional $\frac{2\pi}{3}$ to get the third cube root (Figure 9.5).

Similarly, for each natural number n, the complex n^{th} roots of 1 can be obtained by starting at 1 and successively moving around the unit circle in a counterclockwise direction through angles of $\frac{2\pi}{n}$.

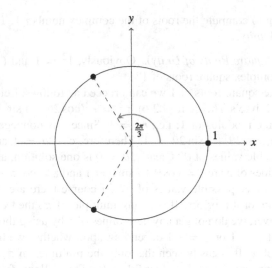

Fig. 9.5 The cube roots of 1

Example 9.2.13 (n^{th} Roots of Unity).   For each natural number n, the complex n^{th} roots of 1 are the numbers 1, $\cos \frac{2\pi}{n} + i \sin \frac{2\pi}{n}$, $\cos \frac{4\pi}{n} + i \sin \frac{4\pi}{n}$, $\cos \frac{6\pi}{n} + i \sin \frac{6\pi}{n}$, $\cos \frac{8\pi}{n} + i \sin \frac{8\pi}{n}$, ..., $\cos \frac{2\pi(n-1)}{n} + i \sin \frac{2\pi(n-1)}{n}$.

Proof. To see this, first note that, for any natural number k,

$$\left(\cos \frac{2\pi k}{n} + i \sin \frac{2\pi k}{n} \right)^n = \cos 2\pi k + i \sin 2\pi k$$

by De Moivre's Theorem (9.2.6). Since $\cos 2\pi k + i \sin 2\pi k = 1$, this shows that each of $\cos \frac{2\pi k}{n} + i \sin \frac{2\pi k}{n}$ is an nth root of unity.

To show that these are the only n^{th} roots of unity, we proceed as follows. Suppose that $z = \cos \theta + i \sin \theta$ and $z^n = 1$. Then $\cos n\theta + i \sin n\theta = 1$, so $\cos n\theta = 1$ and $\sin n\theta = 0$. Thus, $n\theta = 2\pi k$ for some integer k. It follows that $\theta = \frac{2\pi k}{n}$. Taking $k = 0, 1, \ldots, n - 1$ gives the nth roots that we have listed. Taking other values of k gives different values for $\frac{2\pi k}{n}$, but each of them differs from one of the listed values by a multiple of 2π and therefore gives a value for $\cos \theta + i \sin \theta$ that we already have. Thus, the n roots that we listed are all of the nth roots of unity. $\qquad\square$

Roots of other complex numbers can also be computed.

Example 9.2.14. All of the solutions of the equation $z^3 = 1 + i$ can be found as follows.

First note that $|1 + i| = \sqrt{2}$ and the argument of $1 + i$ is $\frac{\pi}{4}$. That is, $1 + i = \sqrt{2}\left(\cos \frac{\pi}{4} + i \sin \frac{\pi}{4}\right)$. Suppose that $z = r(\cos \theta + i \sin \theta)$ and $z^3 = 1 + i$. Then $z^3 = r^3(\cos 3\theta + i \sin 3\theta)$. Therefore $r^3 = \sqrt{2}$, so $r = 2^{\frac{1}{6}}$. Clearly 3θ could be $\frac{\pi}{4}$, in which case θ is $\frac{\pi}{12}$. But also, 3θ could be $\frac{\pi}{4}$ plus any integer multiple of 2π. In particular, $3\theta = \frac{\pi}{4} + 2\pi$ yields $\theta = \frac{3\pi}{4}$ and $3\theta = \frac{\pi}{4} + 4\pi$ yields $\theta = \frac{17\pi}{12}$. This gives three solutions of the equation $z^3 = 1 + i$: $2^{\frac{1}{6}}\left(\cos \frac{\pi}{12} + i \sin \frac{\pi}{12}\right)$, $2^{\frac{1}{6}}\left(\cos \frac{3\pi}{4} + i \sin \frac{3\pi}{4}\right)$, and $2^{\frac{1}{6}}\left(\cos \frac{17\pi}{12} + i \sin \frac{17\pi}{12}\right)$.

There are two different ways of seeing that these three are the only solutions of the equation. One way is to verify that $3\theta = \frac{\pi}{4} + 2\pi k$ for any integer k implies that θ differs from one of $\frac{\pi}{12}$, $\frac{3\pi}{4}$ and $\frac{17\pi}{12}$ by an integer multiple of 2π. Alternately, this follows from the fact that a cubic polynomial has at most three roots (see Theorem 9.3.8 below). $\qquad\square$

9.3 The Fundamental Theorem of Algebra

One reason for introducing complex numbers was to provide a root for the polynomial $x^2 + 1$. There are many other polynomials that do not have any real roots. For example, if $p(x)$ is any polynomial, then the polynomial obtained by writing out $\left(p(x)\right)^2 + 1$ has no real roots, since its value is at least 1 for every value of x.

Does every such polynomial have a complex root? More generally, does every polynomial have a complex root? There is a trivial sense in which the answer to this question is "no," since constant polynomials other than 0 clearly do not have any roots of any kind. For other polynomials, the answer is not so simple. It is a remarkable and very useful fact that every non-constant polynomial with real coefficients, or even with complex coefficients, has a complex root.

The Fundamental Theorem of Algebra 9.3.1. *Every non-constant polynomial with complex coefficients has a complex root.*

There are a number of different proofs of the Fundamental Theorem of Algebra. They all rely on mathematical concepts that we do not develop in this book. We will therefore simply discuss implications of this theorem without proving it.

How many roots can a polynomial have?

Example 9.3.2. The only root of the polynomial $p(z) = z^2 - 6z + 9$ is $z = 3$. This follows from the fact that $p(z) = (z - 3)(z - 3)$. Since the product of two complex numbers is 0 only if at least one of the numbers is 0, the only solution to $p(z) = 0$ is $z = 3$.

In some sense, however, this polynomial has 3 as a "double root"; we'll discuss this a little more below.

To explore the question of the number of roots that a polynomial can have, we need to use division of one polynomial by another. This concept of division is very similar to "long division" of one natural number by another. Actually, we only need a special case of this concept, the case where the polynomial divisor is linear (i.e., has degree 1). We begin with an example.

Example 9.3.3. To divide $z - 3$ into $z^4 + 5z^3 - 2z + 1$, proceed as follows:

$$
\begin{array}{r}
z^3 \ +8z^2 \ +24z \ +70 \\
z-3 \ \overline{)\ z^4 \ +5z^3 \qquad\qquad -2z \quad +1} \\
\underline{z^4 \ -3z^3} \\
8z^3 \qquad\qquad -2z \quad +1 \\
\underline{8z^3 \ -24z^2} \\
24z^2 \ -2z \quad +1 \\
\underline{24z^2 \ -72z} \\
70z \quad +1 \\
\underline{70z \ -210} \\
211
\end{array}
$$

What this calculation shows (like with long division of numbers) is that

$$z^4 + 5z^3 - 2z + 1 = (z - 3)(z^3 + 8z^2 + 24z + 70) + 211$$

The only consequence of the division of one polynomial by another that we need for present purposes is the following.

Theorem 9.3.4. *If r is a complex number and $p(z)$ is a non-constant polynomial with complex coefficients, then there exists a polynomial $q(z)$ and a constant c such that*

$$p(z) = (z - r)q(z) + c$$

Proof. We will proceed by using the Principle of Complete Mathematical Induction (2.2.1) on the degree of the polynomial $p(z)$. Since $p(z)$ is non-constant, the base case of our induction proof is when the degree of $p(z)$ is 1. In other words, $p(z) = az + b$, where a and b are complex numbers and $a \neq 0$. Let r be a complex number. As in Example 9.3.3, we use long division to divide $z - r$ into $p(z)$:

$$
\begin{array}{r}
a \\
z - r \,\overline{\smash{\big)}\, az \quad\; +b} \\
\underline{az \quad -ar} \\
ar + b
\end{array}
$$

This shows that $p(z) = az + b = (z - r) \cdot a + (ar + b)$. Therefore, setting $q(z) = a$ and $c = ar + b$ gives us the desired result when the degree of $p(z)$ is 1. Thus, the base case of the induction is established.

Now assume that the theorem is true for all polynomials of degree less than or equal to n. Using this assumption, we will show that the theorem holds for every polynomial of degree $n + 1$. Let $p(z)$ be a polynomial with complex coefficients with degree $n + 1$. That is,

$$p(z) = a_{n+1}z^{n+1} + a_n z^n + a_{n-1}z^{n-1} + \cdots + a_1 z + a_0$$

where each a_i is a complex number and a_{n+1} is nonzero. Let r be a complex number. Once again we use "long division" to divide $z - r$ into $p(z)$:

$$
\begin{array}{r}
a_{n+1}z^n \\
z - r \,\overline{\smash{\big)}\, a_{n+1}z^{n+1} \qquad\quad +a_n z^n + a_{n-1}z^{n-1} + \cdots + a_1 z + a_0} \\
\underline{a_{n+1}z^{n+1} \qquad -ra_{n+1}z^n} \\
(a_n + ra_{n+1})z^n + a_{n-1}z^{n-1} + \cdots + a_1 z + a_0
\end{array}
$$

To simplify the notation, let $p_n(z) = (a_n + ra_{n+1})z^n + a_{n-1}z^{n-1} + \cdots + a_1 z + a_0$. Then the above gives $p(z) = (z - r)(a_{n+1}z^n) + p_n(z)$. Since $p_n(z)$ is a polynomial of degree less than or equal to n, the induction hypothesis tells us that there exist a polynomial $q_n(z)$ and a constant c such that $p_n(z) = (z - r)q_n(z) + c$. Thus,

$$
\begin{aligned}
p(z) \;=\;& (z - r)(a_{n+1}z^n) + p_n(z) \;=\; (z - r)(a_{n+1}z^n) + (z - r)q_n(z) + c \\
=\;& (z - r)\big(a_{n+1}z^n + q_n(z)\big) + c
\end{aligned}
$$

Therefore, setting $q(z) = a_{n+1}z^n + q_n(z)$ gives us the desired result when the degree of $p(z)$ is $n + 1$. □

Definition 9.3.5. The polynomial $f(z)$ is a *factor* of the polynomial $p(z)$ if there exists a polynomial $q(z)$ such that $p(z) = f(z)q(z)$.

The Factor Theorem 9.3.6. *The complex number r is a root of a polynomial $p(z)$ if and only if $z - r$ is a factor of $p(z)$.*

Proof. If $(z - r)$ is a factor of $p(z)$, then $p(z) = (z - r)q(z)$ implies that $p(r) = (r - r)q(r) = 0 \cdot q(r) = 0$. Conversely, suppose that r is a root of $p(z)$. By Theorem 9.3.4, $p(z) = (z - r)q(z) + c$ for some constant c. Substituting r for z and using the fact that r is a root gives $0 = (r - r)q(r) + c$, so $0 = 0 + c$, from which it follows that $c = 0$. Hence, $p(z) = (z - r)q(z)$ and $z - r$ is a factor of $p(z)$.   □

Example 9.3.7. The complex number $2i$ is a root of the polynomial $iz^3 + z^2 - 4$ (as can be seen by simply substituting $2i$ for z in the expression for the polynomial and noting that the result is 0). It follows from the Factor Theorem (9.3.6) that $z - 2i$ is a factor of the given polynomial. Doing "long division" gives $iz^3 + z^2 - 4 = (z - 2i)(iz^2 - z - 2i)$.

We can use the Factor Theorem to determine the maximum number of roots that a polynomial can have.

Theorem 9.3.8. *A polynomial of degree n has at most n complex roots; if "multiplicities" are counted, it has exactly n roots.*

Proof. Let $p(z)$ be a polynomial of degree n. If n is at least 1, then $p(z)$ has a root, say r_1, by the Fundamental Theorem of Algebra (9.3.1). By the Factor Theorem (9.3.6), there exists a polynomial $q_1(z)$ such that $p(z) = (z - r_1)q_1(z)$. The degree of q_1 is clearly $n - 1$. If $n - 1 > 0$, then $q_1(z)$ has a root, say r_2. It follows from the Factor Theorem that there is a polynomial $q_2(z)$ such that $q_1(z) = (z - r_2)q_2(z)$. The degree of $q_2(z)$ is $n - 2$, and

$$p(z) = (z - r_1)(z - r_2)q_2(z)$$

This process can continue (a formal proof can be given using mathematical induction) until a quotient is simply a constant, say k. Then,

$$p(z) = k(z - r_1)(z - r_2) \cdots (z - r_n)$$

If the r_i are all different, the polynomial will have n roots. If some of the r_i coincide, collecting all the terms where r_i is equal to a given r produces a factor of the form $(z - r)^m$, where m is the number of times that r occurs in the factorization. In this situation, we say that r is a *root of multiplicity m* of the polynomial. Thus, a polynomial of degree n has at most n distinct roots. If the roots are counted according to their multiplicities, then a polynomial of degree n has exactly n roots.   □

9.4 Problems

Basic Exercises

1. Write the following complex numbers in $a + bi$ form, where a and b are real numbers:

   (a) $\left(\frac{1}{\sqrt{2}} + \frac{i}{\sqrt{2}}\right)^{10}$

   (b) $\left(\frac{1}{\sqrt{2}} + \frac{i}{\sqrt{2}}\right)^{106}$

   (c) $\left(\frac{\sqrt{3}}{2} + \frac{i}{2}\right)^{11}$

   (d) $\frac{3+2i}{-1-i}$

   (e) $\frac{3}{\sqrt{2}+i}$

   (f) i^{574}

   (g) i^{575}

   (h) $\frac{1}{i^9}$

2. Show that the real part of $(1 + i)^{10}$ is 0.
3. Find both square roots of each of the following numbers:

   (a) $-i$

   (b) $-15 - 8i$

   [Hint: Suppose $(a + bi)^2 = -15 - 8i$ and compute a and b.]

4. Find all the cube roots of each of the following numbers:

   (a) 2

   (b) $8\sqrt{3} + 8i$

5. (a) Prove that the conjugate of the sum of any two complex numbers is the sum of their conjugates.

   (b) Prove that the conjugate of the product of any two complex numbers is the product of their conjugates.

Interesting Problems

6. Prove the *Quadratic Formula*; i.e., prove that the polynomial $az^2 + bz + c$, where a, b and c are any complex numbers and a is different from 0, has roots

   $$z = \frac{-b \pm \sqrt{b^2 - 4ac}}{2a}.$$

   [Hint: Rewrite the equation as $z^2 + \frac{b}{a}z + \frac{c}{a} = 0$, and use the fact that $\left(z + \frac{b}{2a}\right)^2 = z^2 + \frac{b}{a}z + \frac{b^2}{4a^2}$.]

7. Find all solutions to the equation $iz^2 + 2z + i = 0$.

8. Find a polynomial p with integer coefficients such that $p\left(3 + i\sqrt{7}\right) = 0$.

9. Find all the complex roots of the polynomial $z^6 + z^3 + 1$.

10. Find all the complex roots of the polynomial $z^7 - z$.

11. Find a polynomial whose complex roots are $2 - i$, $2 + i$, 7.

Challenging Problems

12. Find all the complex solutions of $\dfrac{z^3 + 1}{z^3 - 1} = i$.

13. Let p be a polynomial with real coefficients. Prove that the complex conjugate of each root of p is also a root of p.
    [Hint: Use Problem 5 in this chapter.]

14. Show that every non-constant polynomial with real coefficients can be factored into a product of linear (i.e., of degree 1) and quadratic (i.e., of degree 2) polynomials, each of which also has real coefficients.

15. Extend De Moivre's Theorem (9.2.6) to prove that, for negative integers n,

$$\big(r(\cos\theta + i\sin\theta)\big)^n = r^n(\cos n\theta + i\sin n\theta)$$

Chapter 10
Sizes of Infinite Sets

How many natural numbers are there? How many even natural numbers are there? How many odd natural numbers are there? How many rational numbers are there? How many real numbers are there? How many points are there in the plane? How many sets of natural numbers are there? How many different circles are there in the plane? An answer to all these questions could simply be: there are an infinite number of them. But there are more precise answers that can be given; there are, in a sense that we will explain, an infinite number of different size infinities.

10.1 Cardinality

Definition 10.1.1. By a *set* we simply mean any collection of things; the things are called *elements of the set*. (As will be discussed at the end of this chapter, such a general definition of a set is problematic in certain senses.)

For example, the collection of all words on this page is a set. The collection containing the letters a, b, and c is a set: it could be denoted $\{a, b, c\}$. The set of all real numbers greater than 4 could be written:

$$\{x : x > 4\}$$

The fact that something is an element of a set is often denoted using the Greek letter epsilon, $\in$. We write $x \in S$ to represent the fact that x is an element of the set S. For example, if $S = \{x : x > 4\}$, then $17 \in S$.

Definition 10.1.2. If S is a set, a *subset* of S is a set all of whose elements are elements of the set S. The notation $\mathcal{T} \subset S$ is used to signify that $\mathcal{T}$ is a subset of S. The *empty set* is the set that has no elements at all. It is denoted $\emptyset$. The empty set is, by definition, a subset of every set. That is, $\emptyset \subset S$ for every set S.

© Springer Nature Switzerland AG 2018
D. Rosenthal et al., *A Readable Introduction to Real Mathematics*,
Undergraduate Texts in Mathematics, https://doi.org/10.1007/978-3-030-00632-7_10

The *union* of a collection of sets is the set consisting of all elements that occur in at least one of the given sets. The union of sets S and T is denoted $S \cup T$ and similar notation is used for the union of more than two sets.

The *intersection* of a collection of sets is the set consisting of all elements that are in every set in the given collection. The intersection of the sets S and T is denoted $S \cap T$ and similar notation is used for the intersection of more than two sets. If the intersection of two sets is the empty set, the sets are said to be *disjoint*.

How should we define the concept that two sets have the same number of elements? For finite sets, we count the number of elements in each set. When we count the number of elements in a set, we assign the number 1 to one of the elements of the set, then assign the number 2 to another element of the set, then 3 to another element of the set, and so on, until we have counted every element in the set. If the set has n elements, when we finish counting we will have assigned a number in the set $\{1, 2, 3, \ldots, n\}$ to each element of the set and will not have assigned two different numbers to the same element in the set. That is, counting that a set has n elements produces a pairing of the elements of the set $\{1, 2, 3, \ldots, n\}$ with the elements of the set that we are counting. A set whose elements can be paired with the elements of the set $\{1, 2, 3, \ldots, n\}$ is said to have n elements.

More generally, we can say that two sets have the same number of elements if the elements of those two sets can be paired with each other.

Example 10.1.3. Pairs of running shoes are manufactured in a given factory. Each day, some number of pairs is manufactured. Even without knowing how many pairs were manufactured in a given day, we can still conclude that the same number of left shoes was manufactured as the number of right shoes that was manufactured, since they are manufactured in pairs. If, for example, the number of left shoes was determined to be 1012, then it could be concluded that the number of right shoes was also 1012. This could be established as follows: since the set $\{1, 2, 3, \ldots, 1012\}$ can be paired with the set of left shoes, it could also be paired with the set of right shoes, simply by pairing each right shoe to the number assigned to the corresponding left shoe in the pair.

The above discussion suggests the general definition that we shall use. In the following, the phrase "have the same cardinality" is the standard mathematical terminology for what might colloquially be expressed "have the same size."

We will say that the sets S and T "have the same cardinality" if there is a pairing of the elements of S with the elements of T.

We need to precisely define what is meant by a "pairing" of the elements of two sets. This can be specified in terms of functions. A *function* from a set S into a set T is simply an assignment of an element of T to each element of S. For example, if $S = \{a, b, d, e\}$ and $T = \{+, \pi\}$, then one particular function taking S to T is the function f defined by $f(a) = \pi$, $f(b) = \pi$, $f(d) = +$, and $f(e) = \pi$.

Definition 10.1.4. The notation $f : S \to T$ is used to denote a function f taking the set S into the set T; that is, a mapping of each element of S to an element of T. The set S is called the *domain* of the function. The *range* of a function is the set of all its values; that is, the range of $f : S \to T$ is $\{f(s) : s \in S\}$.

Definition 10.1.5. A function $f : S \to T$ is *one-to-one* (or *injective*) if $f(s_1) \neq f(s_2)$ whenever $s_1 \neq s_2$. That is, a function is one-to-one if it does not send two different elements to the same element.

We also require another property that functions may have.

Definition 10.1.6. A function $f : S \to T$ is *onto* (or *surjective*) if for every $t \in T$ there is an $s \in S$ such that $f(s) = t$; that is, the range of f is all of T.

Note that a one-to-one, onto function from a set S onto a set T gives a pairing of the elements of S with those of T.

The formal definition of when sets are to be considered to have the same size can be stated as follows.

Definition 10.1.7. The sets S and T *have the same cardinality* if there is a function $f : S \to T$ that is one-to-one and is onto all of T.

We require the concept of the inverse of a function. If f is a one-to-one function mapping a set S onto a set T, then there is a function mapping T onto S that "sends elements back to where they came from" via f.

Definition 10.1.8. If f is a one-to-one function mapping S onto T, then the *inverse of f*, often denoted f^{-1}, is the function mapping T onto S defined by $f^{-1}(t) = s$ when $f(s) = t$.

With respect to this definition, note that f must be onto for f^{-1} to be defined on all of T. Also, f must be one-to-one; otherwise for some t there will be more than one s for which $f(s) = t$ and therefore $f^{-1}(t)$ will not be determined. If f is a one-to-one function mapping S onto T, then f^{-1} is a one-to-one function mapping T onto S.

Let's consider some examples.

Example 10.1.9. The set of even natural numbers and the set of odd natural numbers have the same cardinality.

Proof. Write the set of even natural numbers as $\mathcal{E} = \{2, 4, \ldots, 2n, \ldots\}$ and the set of odd natural numbers as $O = \{1, 3, \ldots, 2n - 1, \ldots\}$. To satisfy Definition 10.1.7, we need to show that there is a one-to-one function taking $\mathcal{E}$ onto O. Define a function f taking $\mathcal{E} \to O$ by letting $f(k) = k - 1$, for each k in $\mathcal{E}$. To see that this f is one-to-one, simply note that $k_1 - 1 = k_2 - 1$ implies $k_1 = k_2$. Also, f is clearly onto. Thus, the sets $\mathcal{E}$ and O have the same cardinality. $\square$

It is not very surprising that the set of even natural numbers and the set of odd natural numbers have the same cardinalities. The following example is a little more unexpected.

Example 10.1.10. The set of even natural numbers has the same cardinality as the set of all natural numbers.

Proof. This is surprising at first because it seems that the set of even numbers should have half as many elements as the set of all natural numbers. However, it is easy to prove that these sets, $\mathcal{E}$ and $\mathbb{N}$, do have the same cardinality. Simply define the function $f : \mathbb{N} \to \mathcal{E}$ by $f(n) = 2n$. It is easily seen that f is one-to-one: if $f(n_1) = f(n_2)$, then $2n_1 = 2n_2$, so $n_1 = n_2$. The function f is onto since every even number is of the form $2k$ for some natural number k. Therefore, $\mathbb{N}$ and $\mathcal{E}$ have the same cardinality. □

Thus, in the sense of the definition we are using, the subset $\mathcal{E}$ of $\mathbb{N}$ has the same size as the entire set $\mathbb{N}$. This shows that, with respect to cardinality, it is not necessarily the case that "the whole is greater than any of its parts."

Another example showing that "the whole" can have the same cardinality as "one of its parts" is the following.

Example 10.1.11. The set of natural numbers and the set of nonnegative integers have the same cardinality.

Proof. The set of natural numbers is $\mathbb{N} = \{1, 2, 3, \ldots\}$. Let S denote the set $\{0, 1, 2, 3, \ldots\}$ of nonnegative integers. We want to construct a one-to-one function f taking S onto $\mathbb{N}$. We can simply define f by $f(n) = n + 1$, for each n in S. Clearly f maps S onto $\mathbb{N}$. Also, $f(n_1) = f(n_2)$ implies $n_1 + 1 = n_2 + 1$, which gives $n_1 = n_2$. That is, f does not send two different integers to the same natural number, so f is one-to-one. Therefore, $\mathbb{N}$ and S have the same cardinality. □

The following notation is useful.

Notation 10.1.12. We use the notation $|S| = |\mathcal{T}|$ to mean that S and $\mathcal{T}$ have the same cardinality.

Therefore, as shown above, $|O| = |\mathcal{E}| = |\mathbb{N}|$.

How does the size of the set of all positive rational numbers, which we will denote by $\mathbb{Q}^+$, compare to the size of the set of natural numbers? The subset of $\mathbb{Q}^+$ consisting of those rational numbers with numerator 1 can obviously be paired with $\mathbb{N}$: simply pair $\frac{1}{n}$ with n, for each n in $\mathbb{N}$. But then there are all the rational numbers with numerator 2, and with numerator 3, and so on. It seems that there are many more positive rational numbers than there are natural numbers. However, we now prove that $|\mathbb{N}| = |\mathbb{Q}^+|$.

Theorem 10.1.13. *The set of natural numbers and the set of positive rational numbers have the same cardinality.*

Proof. To prove this theorem, we first describe a way of displaying all the positive rational numbers. We imagine writing all the rational numbers with numerator 1 in one line, and then, underneath that, the rational numbers with numerator 2 in a line, and under that the rational numbers with numerator 3 in a line, and so on. That is, we consider the following array:

$$\frac{1}{1} \quad \frac{1}{2} \quad \frac{1}{3} \quad \frac{1}{4} \quad \frac{1}{5} \quad \frac{1}{6} \quad \frac{1}{7} \quad \cdots$$

$$\frac{2}{1} \quad \frac{2}{2} \quad \frac{2}{3} \quad \frac{2}{4} \quad \frac{2}{5} \quad \frac{2}{6} \quad \frac{2}{7} \quad \cdots$$

$$\frac{3}{1} \quad \frac{3}{2} \quad \frac{3}{3} \quad \frac{3}{4} \quad \frac{3}{5} \quad \frac{3}{6} \quad \frac{3}{7} \quad \cdots$$

$$\frac{4}{1} \quad \frac{4}{2} \quad \frac{4}{3} \quad \frac{4}{4} \quad \frac{4}{5} \quad \frac{4}{6} \quad \frac{4}{7} \quad \cdots$$

$$\vdots \quad \vdots \quad \vdots \quad \vdots \quad \vdots \quad \vdots \quad \vdots$$

Imagining the positive rational numbers arranged as above, we can show that the natural numbers can be paired with them. That is, we will define a one-to-one function f taking $\mathbb{N}$ onto $\mathbb{Q}^+$. As we define the function, you should keep looking back at the array to see the pattern that we are using.

Define $f(1) = \frac{1}{1}$ and $f(2) = \frac{1}{2}$. (We can't continue by $f(3) = \frac{1}{3}$, $f(4) = \frac{1}{4}, \ldots$, for then f would only map onto those rational numbers with numerator 1.) Define $f(3) = \frac{2}{1}$ and $f(4) = \frac{3}{1}$. We can't just keep going down in our array; we must include the numbers above as well. We do not include $\frac{2}{2}$ however, since $\frac{2}{2} = \frac{1}{1}$, which is already paired with 1. Thus, we let $f(5) = \frac{1}{3}$, $f(6) = \frac{1}{4}$, $f(7) = \frac{2}{3}$, $f(8) = \frac{3}{2}$, $f(9) = \frac{4}{1}$, and $f(10) = \frac{5}{1}$. We do not consider $\frac{4}{2}$, since $\frac{4}{2} = \frac{2}{1}$, and we do not consider $\frac{3}{3} = \frac{1}{1}$ or $\frac{2}{4} = \frac{1}{2}$. Thus, $f(11)$ is defined to be $\frac{1}{5}$ and $f(12) = \frac{1}{6}$. It is apparent that a pairing of the natural numbers and the positive rational numbers is indicated by continuing to label rational numbers with natural numbers in this manner, "zigzagging," you might say, through the above array. Therefore, $|\mathbb{Q}^+| = |\mathbb{N}|$.   □

10.2   Countable Sets and Uncountable Sets

You may be wondering whether or not every infinite set can be paired with the set of natural numbers. If the elements of a set can be paired with the natural numbers, then the elements can be listed in a sequence. For example, if we let s_1 be the element of the set corresponding to the natural number 1, s_2 be the element of the set corresponding to the natural number 2, s_3 to 3, and so on, then the set could be displayed:

$$\{s_1, s_2, s_3, \ldots\}$$

Pairing the elements of a set with the set of natural numbers is, in a sense, "counting the elements of the set."

Definition 10.2.1. A set is *countable* (sometimes called *denumerable*, or *enumerable*) if it is either finite or has the same cardinality as the set of natural numbers. A set is said to be *uncountable* if it is not countable.

Definition 10.2.2. For a and b real numbers with $a \leq b$, the *closed interval from a to b* is the set of all real numbers between a and b, including a and b. It is denoted $[a, b]$. That is, $[a, b] = \{x : a \leq x \leq b\}$.

One example of an uncountable set is the following.

Theorem 10.2.3. *The closed interval* $[0, 1]$ *is uncountable.*

Proof. We must prove that there is no way of pairing the set of natural numbers with the interval $[0, 1]$. To establish this, we will show that every pairing of natural numbers with elements of $[0, 1]$ fails to include some members of $[0, 1]$. In other words, we will show that there does not exist any function that maps $\mathbb{N}$ onto $[0, 1]$.

We use the fact that the elements of $[0, 1]$ can be written as infinite decimals; that is, in the form $.c_1c_2c_3\ldots$, where each c_i is a digit between 0 and 9. (This fact will be formally established in Chapter 13 of this book (see Theorem 13.6.3).) Some numbers have two different such representations. For example, $.9999\ldots$ is the same number as $1.0000\ldots$, and $.19999\ldots$ is the same number as $.20000\ldots$ (see Section 13.6). For the rest of this proof, let us agree that we choose the representation involving an infinite string of 9's rather than the representation involving an infinite string of 0's for all numbers that have two different representations.

Suppose that f is any function taking $\mathbb{N}$ to $[0, 1]$. To prove that f cannot be onto, we imagine writing out all the values of f in a list, as follows:

$$f(1) = .a_{11}a_{12}a_{13}a_{14}a_{15}\ldots$$

$$f(2) = .a_{21}a_{22}a_{23}a_{24}a_{25}\ldots$$

$$f(3) = .a_{31}a_{32}a_{33}a_{34}a_{35}\ldots$$

$$f(4) = .a_{41}a_{42}a_{43}a_{44}a_{45}\ldots$$

$$f(5) = .a_{51}a_{52}a_{53}a_{54}a_{55}\ldots$$

$$\vdots$$

We now construct a number in $[0, 1]$ that is not in the range of the function f. We do that by showing how to choose digits b_j so that the number $x = .b_1b_2b_3b_4\ldots$ is not in the range of f. Begin by choosing $b_1 = 3$ if $a_{11} \neq 3$ and $b_1 = 4$ if $a_{11} = 3$. Then choose $b_2 = 3$ if $a_{22} \neq 3$ and $b_2 = 4$ if $a_{22} = 3$. We continue in this manner, choosing $b_j = 3$ if $a_{jj} \neq 3$ and $b_j = 4$ if $a_{jj} = 3$, for every natural number j. The number x that is so constructed has a unique decimal representation (since there are no 9's or 0's in its representation) and differs from $f(j)$ in its jth digit. Therefore, $f(j) \neq x$ for all j, so x is not in the range of f. That is, we have proven that there is no function (one-to-one or otherwise) taking $\mathbb{N}$ onto $[0, 1]$, so we conclude that $[0, 1]$ has cardinality different from that of $\mathbb{N}$.   □

Of course, any given function f in the above proof could be modified so as to produce a function whose range does include the specific number x that we constructed in the course of the proof. For example, given any such function f, define the function $g : \mathbb{N} \to [0, 1]$ by defining $g(1) = x$ and $g(n) = f(n-1)$, for $n \geq 2$. The range of g includes x and also includes the range of f. However, g does not map $\mathbb{N}$ onto $[0, 1]$, for the above proof could be used to produce a different x that is not in the range of g.

How does the cardinality of other closed intervals compare to that of $[0, 1]$?

Theorem 10.2.4. *If a and b are real numbers and $a < b$, then $[a, b]$ and $[0, 1]$ have the same cardinality.*

Proof. The theorem will be established if we construct a function $f : [0, 1] \to [a, b]$ that is one-to-one and onto. That is easy to do. Simply define f by $f(x) = a + (b-a)x$. Then $f(0) = a$ and $f(1) = b$. If x is in $[0, 1]$, then $a + (b-a)x$ is greater than or equal to a and less than or equal to b, so f takes $[0, 1]$ into $[a, b]$. To show that f is onto, let y be any element of $[a, b]$. Let $x = \frac{y-a}{b-a}$. Then $x \in [0, 1]$ and $f(x) = y$. This shows that f is onto. To show that f is one-to-one, assume that $a + (b-a)x_1 = a + (b-a)x_2$. Subtracting a from both sides of this equation and then dividing both sides by $b - a$ yields $x_1 = x_2$. This proves that f is one-to-one. Thus, f is a pairing of the elements of $[0, 1]$ with the elements of $[a, b]$, so $\left|[0, 1]\right| = \left|[a, b]\right|$. $\qquad \square$

There are other intervals that frequently arise in mathematics.

Definition 10.2.5. If a and b are real numbers and $a < b$, then the *open interval between a and b*, denoted (a, b), is defined by

$$(a, b) = \{x : a < x < b\}$$

The *half-open intervals* are defined by

$$(a, b] = \{x : a < x \leq b\} \text{ and } [a, b) = \{x : a \leq x < b\}$$

How does the size of a half-open interval compare to the size of the corresponding closed interval?

Theorem 10.2.6. *The intervals $[0, 1]$ and $(0, 1]$ have the same cardinality.*

Proof. We want to construct a one-to-one function f taking $[0, 1]$ onto $(0, 1]$. We will define $f(x) = x$ for most x in $[0, 1]$, but we need to make a place for 0 to go to in the half-open interval. For each natural number n, the rational number $\frac{1}{n}$ is in both intervals. Define f on those numbers by $f\left(\frac{1}{n}\right) = \frac{1}{n+1}$ for $n \in \mathbb{N}$. In particular, $f(1) = \frac{1}{2}$. Note that the number 1, which is in $(0, 1]$, is not in the range of f as defined so far. We define $f(0)$ to be 1. We define f on the rest of $[0, 1]$ by $f(x) = x$. That is, $f(x) = x$ for those x other than 0 that are not of the form $\frac{1}{n}$ with n a natural number. It is straightforward to check that we have constructed a one-to-one function mapping $[0, 1]$ onto $(0, 1]$. $\qquad \square$

Suppose that $|S| = |\mathcal{T}|$ and $|\mathcal{T}| = |\mathcal{U}|$; must $|S| = |\mathcal{U}|$? If this was not the case, we would be using the "equals" sign in a very peculiar way.

Theorem 10.2.7. *If $|S| = |\mathcal{T}|$ and $|\mathcal{T}| = |\mathcal{U}|$, then $|S| = |\mathcal{U}|$.*

Proof. By hypothesis, there exist one-to-one functions f and g mapping S onto $\mathcal{T}$ and $\mathcal{T}$ onto $\mathcal{U}$, respectively. That is, $f : S \rightarrow \mathcal{T}$ and $g : \mathcal{T} \rightarrow \mathcal{U}$. Let $h = g \circ f$ be the composition of g and f. In other words, h is the function defined on S by $h(s) = g(f(s))$. We must show that h is a one-to-one function taking S onto $\mathcal{U}$. Let u be any element of $\mathcal{U}$. Since g is onto, there exists a t in $\mathcal{T}$ such that $g(t) = u$. Since f is onto, there is an s in S such that $f(s) = t$. Then $h(s) = g(f(s)) = g(t) = u$. Thus, h is onto.

To see that h is one-to-one, suppose that $h(s_1) = h(s_2)$; we must show that $s_1 = s_2$. Now $g(f(s_1)) = g(f(s_2))$, so $f(s_1) = f(s_2)$ since g is one-to-one. But f is also one-to-one, and so $s_1 = s_2$. We have shown that h is one-to-one and onto, from which it follows that $|S| = |\mathcal{U}|$. $\square$

Theorem 10.2.8. *If a, b, c, and d are real numbers with $a < b$ and $c < d$, then the half-open intervals $(a, b]$ and $(c, d]$ have the same cardinality.*

Proof. The function f defined by $f(x) = a + (b - a)x$ is a one-to-one function mapping $(0, 1]$ onto $(a, b]$, as can be seen by a proof almost exactly the same as that in Theorem 10.2.4. Hence, $|(0, 1]| = |(a, b]|$. Similarly the function g defined by $g(x) = c + (d - c)x$ is a one-to-one function mapping $(0, 1]$ onto $(c, d]$, so $|(0, 1]| = |(c, d]|$. It follows from Theorem 10.2.7 that $|(a, b]| = |(c, d]|$. $\square$

Are there more nonnegative real numbers than there are real numbers in $[0, 1]$? The, perhaps surprising, answer is "no."

Theorem 10.2.9. *The cardinality of the set of nonnegative real numbers is the same as the cardinality of the unit interval $[0, 1]$.*

Proof. We begin by showing that the set $S = \{x : x \geq 1\}$ has the same cardinality as $(0, 1]$. Note that the function f defined by $f(x) = \frac{1}{x}$ maps S into $(0, 1]$; for if $x \geq 1$, then $\frac{1}{x} \leq 1$. Also, f maps S onto $(0, 1]$; for if $y \in (0, 1]$, then $\frac{1}{y} \geq 1$ and $f\left(\frac{1}{y}\right) = y$. To see that f is one-to-one, suppose that $f(x_1) = f(x_2)$. Then $\frac{1}{x_1} = \frac{1}{x_2}$, so $x_1 = x_2$. Hence, f is one-to-one and onto, and it follows that $|S| = |(0, 1]|$.

Now let $\mathcal{T} = \{x : x \geq 0\}$. Define the function g by $g(x) = x - 1$. Then g is obviously a one-to-one function mapping S onto $\mathcal{T}$. Hence $|\mathcal{T}| = |S|$. Therefore, by Theorem 10.2.7, $|\mathcal{T}| = |(0, 1]|$. But, by Theorem 10.2.6, $|[0, 1]| = |(0, 1]|$. It follows that $|\mathcal{T}| = |[0, 1]|$. $\square$

Must the union of two countable sets be countable? A much stronger result is true.

Theorem 10.2.10. *The union of a countable number of countable sets is countable.*

Proof. This can be proven using ideas similar to those used in the proof of the fact that the set of positive rational numbers is countable (see Theorem 10.1.13). Recall that "countable" means either finite or having the same cardinality as $\mathbb{N}$ (Definition 10.2.1). We will prove this theorem for the cases where all the sets are infinite; you should be able to see how to modify the proof if some or all of the cardinalities are finite.

Suppose, then, that we have a countable collection $\{S_1, S_2, S_3, \ldots\}$ of sets, each of which is itself countably infinite. By pairing the elements of S_i with the elements of $\mathbb{N}$, we label the elements of S_i so that $S_i = \{a_{i1}, a_{i2}, a_{i3}, \ldots\}$. We display the sets in the following array:

$$S_1 = \{a_{11}, a_{12}, a_{13}, a_{14}, a_{15}, a_{16}, a_{17}, \ldots\}$$

$$S_2 = \{a_{21}, a_{22}, a_{23}, a_{24}, a_{25}, a_{26}, a_{27}, \ldots\}$$

$$S_3 = \{a_{31}, a_{32}, a_{33}, a_{34}, a_{35}, a_{36}, a_{37}, \ldots\}$$

$$S_4 = \{a_{41}, a_{42}, a_{43}, a_{44}, a_{45}, a_{46}, a_{47}, \ldots\}$$

$$\vdots$$

Let S denote the union of the S_i's. To show that S is countable, we show that we can list all of its elements. Proceed as follows. First, list a_{11} and then a_{12}. Then consider a_{21}. It is possible that a_{21} is one of a_{11} or a_{12}, in which case we do not, of course, list it again. If, however, a_{21} is neither a_{11} nor a_{12}, we list it next. Then look at a_{31}; if it is not yet listed, list it next. Then go back up to a_{22}, then a_{13}, and so on. In this way, we "zigzag" through the entire array (as we did in the proof of Theorem 10.1.13) and list all the elements of S. It follows that S is countable. □

10.3 Comparing Cardinalities

When two sets have different cardinalities, the question arises of whether we can say that one set has cardinality that is less than the cardinality of the other set. What should we mean by saying that the cardinality of one set is less than that of another set? It is easiest to begin with a definition of "less than or equal to," instead of "less than," for cardinalities.

Definition 10.3.1. If S and $\mathcal{T}$ are sets, we say that S *has cardinality less than or equal to the cardinality of* $\mathcal{T}$, and write $|S| \leq |\mathcal{T}|$, if there is a subset $\mathcal{T}_0$ of $\mathcal{T}$ such that $|S| = |\mathcal{T}_0|$.

This is equivalent to saying that there is a one-to-one function mapping S into (not necessarily onto) $\mathcal{T}$. For if f is a one-to-one function mapping S onto $\mathcal{T}_0$, we can regard f as a function taking S into $\mathcal{T}$. Conversely, if f is a one-to-one function mapping S into $\mathcal{T}$, and if $\mathcal{T}_0$ is the range of f, then f gives a pairing of S and $\mathcal{T}_0$.

Example 10.3.2. The function $f : \mathbb{N} \to [0, 1]$ defined by $f(n) = \frac{1}{n}$ establishes that $|\mathbb{N}| \le |[0, 1]|$, since f is one-to-one.

Note that $|S_0| \le |S|$ whenever S_0 is a subset of S, since the function $f : S_0 \to S$ defined by $f(s) = s$, for each s in S_0, is clearly one-to-one.

We have defined "$\le$" for cardinalities; how should we define "$<$"? The following definition is very natural.

Definition 10.3.3. We say that S *has cardinality less than that of* $\mathcal{T}$, *and write* $|S| < |\mathcal{T}|$, *if* $|S| \le |\mathcal{T}|$ *and* $|S| \ne |\mathcal{T}|$.

Example 10.3.4. If $\mathbb{N}$ is the set of natural numbers and $[0, 1]$ is the unit interval, then $|\mathbb{N}| < |[0, 1]|$.

Proof. By Example 10.3.2, $|\mathbb{N}| \le |[0, 1]|$, and, by Theorem 10.2.3, $|\mathbb{N}| \ne |[0, 1]|$, so the result follows.   □

Thus, in the sense of the definitions we are using, there are more real numbers in the interval $[0, 1]$ than there are natural numbers.

There is a question that immediately arises from the definition of "less than or equal to" for cardinalities: If S and $\mathcal{T}$ are sets such that $|S| \le |\mathcal{T}|$ and $|\mathcal{T}| \le |S|$, must $|S| = |\mathcal{T}|$? The language suggests that this question should have an affirmative answer, but that language doesn't prove anything. What does this question come down to? We are given the fact that $|S| \le |\mathcal{T}|$. That is equivalent to the existence of a one-to-one function $f : S \to \mathcal{T}$. Similarly, $|\mathcal{T}| \le |S|$ implies that there is a one-to-one function $g : \mathcal{T} \to S$. To say that $|S| = |\mathcal{T}|$ is equivalent to saying that there exists a function $h : S \to \mathcal{T}$ that is both one-to-one and onto. The question, therefore, is whether we can show the existence of such a function h from the existence of the functions f and g.

In addition to being important in justifying the above terminology, the following theorem is often very useful in proving that given sets have the same cardinalities.

The Cantor–Bernstein Theorem 10.3.5. *If S and $\mathcal{T}$ are sets such that $|S| \le |\mathcal{T}|$ and $|\mathcal{T}| \le |S|$, then $|S| = |\mathcal{T}|$.*

Proof. The hypotheses imply that there exist one-to-one functions $f : S \to \mathcal{T}$ and $g : \mathcal{T} \to S$; these functions may or may not be onto. We must construct a one-to-one function h that takes S onto $\mathcal{T}$. To do this, we will break S up into three subsets and then define h to be the function f on two of those subsets and the function g^{-1} on the third subset.

Consider any element s of S. Such an s may or may not be in the range of g. If it is in the range of g, then there is exactly one element t_0 in $\mathcal{T}$ such that $g(t_0) = s$, since g is one-to-one. Call such an element t_0 the "immediate ancestor" of s. Similarly, if t is in $\mathcal{T}$ and $f(s_0) = t$ for some s_0 in S, we say that s_0 is the "immediate ancestor" of t. Thus, elements of S have immediate ancestors in $\mathcal{T}$ if they are in the range of g, and elements of $\mathcal{T}$ have immediate ancestors in S if they are in the range of f. Some elements may not have any immediate ancestors.

We will say that an immediate ancestor of an immediate ancestor of an element s in S is an "ancestor" of the element s. That is, if s in S has an immediate ancestor t_0 in T and t_0 has an immediate ancestor s_0 in S, then s_0 is an ancestor of s. Similarly, if t_1 in T has an immediate ancestor s_1 in S and s_1 has an immediate ancestor t_2 in T, we say that t_2 is an ancestor of t_1. We continue backwards whenever possible. In other words, we start with a given element and then keep on finding immediate ancestors unless and until we reach an element that does not have an immediate ancestor. All the ancestors in such a chain of immediate ancestors are called ancestors of the original element of S or T.

For each element that we start with, there are three possibilities. One possibility is that the element has an ancestor and that every ancestor of the element also has an ancestor. That is, it could be that we can keep on going back and back and back indefinitely in the ancestry of a given element. Let S_∞ denote the set of all those elements s in S for which we can keep on finding ancestors without stopping. Similarly, let T_∞ denote the set of all t in T for which we can keep on finding ancestors without stopping. (It might be noted that it is possible that we can keep on finding ancestors indefinitely, but nonetheless there are only a finite number of distinct ancestors. For example, it would be possible that, for some s in S and t in T, $f(s) = t$ and $g(t) = s$. Then the immediate ancestor of s would be t, the immediate ancestor of t would be s, the immediate ancestor of s would be t, and so on. Thus, there would be no stopping the process of finding ancestors, in spite of the fact that each of s and t has only two distinct ancestors, s and t. In this situation, $s \in S_\infty$ and $t \in T_\infty$.)

Those elements of S and T that are not in either of S_∞ or T_∞ have what might be called "ultimate ancestors." That is, since the chain of ancestors comes to a stop, there is a most distant ancestor. Of course, one possibility is that the element has no ancestors at all, in which case we say that element is its own ultimate ancestor. The ultimate ancestor of any given element is either in S or in T. Let S_S denote the set of all elements of S whose ultimate ancestor is in S and let S_T denote the set of all elements of S whose ultimate ancestor is in T. Similarly, let T_S and T_T denote the sets of elements of T whose ultimate ancestors are in S and T, respectively.

Thus, we have divided S into three subsets: S_∞, S_S, and S_T. Every element of S is in exactly one of those subsets. Similarly, every element of T is in exactly one of the subsets T_∞, T_S, or T_T. (Of course, some of the subsets may be empty.)

We can now define the function h. For s in S, we define $h(s)$ to be $f(s)$ if s is in either S_∞ or S_S, and we define $h(s)$ to be $g^{-1}(s)$ if s is in S_T. Note that $g^{-1}(s)$ is defined for all $s \in S_T$ since all the elements of S_T have immediate ancestors in T. We will show that h is a one-to-one function taking S onto T.

Let's first show that h is one-to-one. Suppose that $h(s_1) = h(s_2)$ for s_1 and s_2 in S. We must show that $s_1 = s_2$. If both of s_1 and s_2 are in the union of S_∞ and S_S, then $h(s_1) = f(s_1)$ and $h(s_2) = f(s_2)$. Therefore, $f(s_1) = f(s_2)$. Since f is one-to-one, it follows that $s_1 = s_2$ in this case. Similarly, if both of s_1 and s_2 are in S_T, then $h(s_1) = g^{-1}(s_1)$ and $h(s_2) = g^{-1}(s_2)$. Therefore, $g^{-1}(s_1) = g^{-1}(s_2)$. Applying g to both sides of this equation gives $s_1 = s_2$ in this case.

One case remains: the case where one of s_1 and s_2 is in the union of S_S and S_∞ and the other is in S_T. Suppose that $s_1 \in S_\infty \cup S_S$ and $s_2 \in S_T$. Then $h(s_1) = f(s_1)$ and $h(s_2) = g^{-1}(s_2)$. Therefore, $f(s_1) = g^{-1}(s_2)$. We show that this case cannot arise. If $f(s_1) = g^{-1}(s_2)$, then s_1 is an immediate ancestor of $g^{-1}(s_2)$. Thus, s_1 is an ancestor of s_2. But s_2 is in S_T, so it has an ultimate ancestor in T. Since s_1 is an ancestor of s_2, the ultimate ancestor of s_2 is the ultimate ancestor of s_1. But s_1 being in $S_\infty \cup S_S$ implies that s_1 either has no ultimate ancestor or has an ultimate ancestor in S. This is inconsistent with having an ultimate ancestor in T, so this case does not arise.

We have proven that the function h that we constructed is one-to-one. It remains to be shown that h maps S onto T.

Each t in T is in one of T_S, T_∞, or T_T. We must show that, wherever t lies, there is an s in S such that $h(s) = t$. Suppose first that $t \in T_S$. Since t has an ultimate ancestor in S, it follows that t is in the range of f, so we can consider $f^{-1}(t)$. The ancestors of $f^{-1}(t)$ are also ancestors of t, from which it follows that the ultimate ancestor of $f^{-1}(t)$ is in S. That is, $f^{-1}(t)$ is in S_S. The function h is defined to be f on S_S, so $h(f^{-1}(t)) = f(f^{-1}(t)) = t$. This shows that the range of h contains every element of T_S.

Now consider any t in T_∞. Such a t has an immediate ancestor in S, $f^{-1}(t)$. Since the ancestors of $f^{-1}(t)$ are also ancestors of t, $f^{-1}(t)$ has no ultimate ancestor. That is, $f^{-1}(t)$ is in S_∞. The function h was defined to be the function f on S_∞, so $h(f^{-1}(t)) = f(f^{-1}(t)) = t$. This proves that the range of h contains T_∞.

All that remains to be shown is that the range of h includes T_T. Suppose, then, that t is in T_T. Let $s = g(t)$. Then t is the immediate ancestor of s. Thus, the ultimate ancestor of t is the ultimate ancestor of s. Since the ultimate ancestor of t is in T, the ultimate ancestor of s is in T. In other words, s is in S_T. On elements of S_T, h is defined to be g^{-1}. Thus, $h(s) = g^{-1}(s)$ and, since $s = g(t)$, $h(s) = g^{-1}(g(t)) = t$. This establishes that the range of h includes T_T.

We have therefore shown that, for every t in T, whatever subset of T contains t, there is an s in S such that $h(s) = t$. This proves that h is onto.

Therefore, h is a one-to-one function mapping S onto T, and we conclude that $|S| = |T|$. $\square$

Corollary 10.3.6. *If S is a subset of T and there exists a function $f : T \to S$ that is one-to-one, then S and T have the same cardinality.*

Proof. Since S is a subset of T, $|S| \le |T|$. Since there is a one-to-one function mapping T into S, $|T| \le |S|$. Then, by the Cantor–Bernstein Theorem (10.3.5), $|S| = |T|$. $\square$

The Cantor–Bernstein Theorem can often be used to simplify proofs that given sets have the same cardinalities.

Theorem 10.3.7. *If $a < b$, then $\left|[a, b]\right| = \left|(a, b)\right| = \left|(a, b]\right| = \left|[a, b)\right|$.*

Proof. Clearly, $|(a, b)| \leq |[a, b]|$. Note that $\left[a + \frac{b-a}{3}, b - \frac{b-a}{3}\right]$ is contained in (a, b), so $\left|\left[a + \frac{b-a}{3}, b - \frac{b-a}{3}\right]\right| \leq |(a, b)|$. But, by Theorem 10.2.4, $|[a, b]| = \left|\left[a + \frac{b-a}{3}, b - \frac{b-a}{3}\right]\right|$. Therefore, $|[a, b]| \leq |(a, b)|$. So, by the Cantor–Bernstein Theorem (10.3.5), $|[a, b]| = |(a, b)|$.

The proofs for the half-open intervals are almost exactly the same as the above proof for the open interval. □

What is the cardinality of the set of all real numbers?

Theorem 10.3.8. *The cardinality of the set of all real numbers is the same as the cardinality of the unit interval* $[0, 1]$.

Proof. Let $\mathbb{R}$ denote the set of all real numbers. We will "patch together" some of the results that we have already proven to show that $|\mathbb{R}| \leq |[0, 1]|$.

As we have seen, the set of nonnegative real numbers has the same cardinality as $[0, 1]$ (see Theorem 10.2.9). Thus, there exists a one-to-one function f mapping the set of nonnegative real numbers onto $[0, 1]$. The set of negative real numbers obviously has the same cardinality as the set of positive real numbers, as can be seen by using the mapping that takes x to $-x$. The positive real numbers can be mapped in a one-to-one way into $[0, 1]$. Since $|[0, 1]| = |[3, 4]|$ (Theorem 10.2.4), it follows that the positive real numbers can be mapped in a one-to-one way into $[3, 4]$. Then, using the equivalence of the positive and negative real numbers, we conclude that there is a function g mapping the negative real numbers into $[3, 4]$. We now define a function h mapping $\mathbb{R}$ into $[0, 1] \cup [3, 4]$ by letting h be f on the nonnegative numbers and g on the negative numbers. Then h is a one-to-one function mapping $\mathbb{R}$ into a subset of $[0, 1] \cup [3, 4]$, which is a subset of $[0, 4]$. It follows that $|\mathbb{R}| \leq |[0, 4]|$. On the other hand, $[0, 4]$ is a subset of $\mathbb{R}$, so $|[0, 4]| \leq |\mathbb{R}|$, and, by the Cantor–Bernstein Theorem (10.3.5), $|\mathbb{R}| = |[0, 4]|$. Since $|[0, 4]| = |[0, 1]|$ (Theorem 10.2.4), the theorem follows. □

There is a theorem that can often be used to provide very easy proofs that sets are countable. The next several results form the basis for that theorem and are useful in other contexts as well.

Theorem 10.3.9. *A subset of a countable set is countable.*

Proof. Let S be a countable set. If S is finite, then the result is clear. If S is infinite, then there exists a one-to-one function, say f, mapping the set of natural numbers onto S. Thus, the elements of S can be listed in a sequence, $(f(1), f(2), f(3), f(4), \dots)$. If S_0 is a subset of S, then the elements of S_0 correspond to some of the elements in the sequence. Therefore, the elements of S_0 can also be listed in a sequence, and hence S_0 is either finite or has the same cardinality as $\mathbb{N}$. □

Corollary 10.3.10. *If S is any set and there exists a one-to-one function mapping S into the set of natural numbers, then S is countable.*

Proof. Let f be a one-to-one function taking S into $\mathbb{N}$. The range of f is some subset $\mathcal{T}$ of $\mathbb{N}$. Since f is a one-to-one function taking S onto $\mathcal{T}$, it follows that $|S| = |\mathcal{T}|$. By the previous theorem, $\mathcal{T}$ is countable, and therefore so is S.   □

Definition 10.3.11. A *finite sequence* of elements of a set S is an ordered collection of elements of S of the form $(s_1, s_2, s_3, \ldots, s_k)$.

For example, one finite sequence of rational numbers is $\left(-\frac{1}{2}, -7, \frac{22}{7}, 0\right)$.

Theorem 10.3.12. *The set of all finite sequences of natural numbers is countable.*

Proof. Let S denote the set of all finite sequences of natural numbers. By the above corollary (10.3.10), it suffices to show that there is a one-to-one function g mapping S into $\mathbb{N}$. Here is a description of one such function. We define the value of g at each given finite sequence of natural numbers to be the number whose digits are 1's and 0's, determined as follows: begin with the number of 1's equal to the first number in the given finite sequence, follow that by a 0, then follow that by the number of 1's equal to the second number in the sequence, then another 0, then the number of 1's corresponding to the third number in the sequence, then a 0, and so on, ending with the number of 1's corresponding to the last number in the sequence. For example,

$$g\big((2, 3, 7)\big) = 11011101111111$$

and

$$g\big((5, 1)\big) = 1111101$$

The function g is one-to-one since the unique sequence corresponding to any number in the range of g can be recovered by using the definition of g. For example, the number 111101011111101111111101 corresponds to the sequence $(4, 1, 6, 8, 1)$. Since g is one-to-one and maps S into $\mathbb{N}$, S is countable.   □

Corollary 10.3.13. *If $\mathcal{L}$ is any countable set, then the set of all finite sequences of elements of $\mathcal{L}$ is countable.*

Proof. This follows easily from the above theorem. By hypothesis, there exists a one-to-one function f mapping $\mathcal{L}$ into $\mathbb{N}$. Then, a one-to-one function F mapping sequences of elements of $\mathcal{L}$ into sequences of elements of $\mathbb{N}$ can be obtained by defining

$$F(a_1, a_2, a_3, \ldots, a_k) = \big(f(a_1), f(a_2), f(a_3), \ldots, f(a_k)\big)$$

Thus, the previous theorem implies the corollary.   □

The following definition will be useful.

Definition 10.3.14. Let $\mathcal{L}$ and $\mathcal{T}$ be any sets. We will say that $\mathcal{T}$ *can be labeled by* $\mathcal{L}$ if there is a one-to-one function from $\mathcal{T}$ to the set of finite sequences of elements of $\mathcal{L}$. In other words, $\mathcal{L}$ can label $\mathcal{T}$ if there is a way of assigning to each element of $\mathcal{T}$ a finite sequence of elements of $\mathcal{L}$ so that no two finite sequences correspond to the same element of $\mathcal{T}$.

Example 10.3.15. The set $\mathbb{N}$ of natural numbers can be labeled by the set of digits $\mathcal{L} = \{0, 1, 2, 3, 4, 5, 6, 7, 8, 9\}$. A given natural number can be labeled by the sequence from $\mathcal{L}$ consisting of its digits in the order in which they occur. For example, 79288 could be labeled by the sequence $(7, 9, 2, 8, 8)$.

Example 10.3.16. The set $\mathbb{Q}$ of rational numbers can be labeled by the set

$$\mathcal{L} = \{0, 1, 2, 3, 4, 5, 6, 7, 8, 9, -, /\}$$

To label a given rational number, first express it in lowest terms and then assign to it the sequence from $\mathcal{L}$ consisting of the minus sign if the number is negative, and then listing the digits of the numerator in the order in which they occur, the $/$, and then the digits of the denominator in their order. For example, $\frac{7}{125}$ would be labeled by the sequence $(7, /, 1, 2, 5)$ and $-\frac{29}{38}$ would be labeled by the sequence $(-, 2, 9, /, 3, 8)$.

The following theorem is useful in many situations. It is a slight variant of the "Typewriter Principle" that was developed by the mathematician Bjorn Poonen.

The Enumeration Principle 10.3.17. *Every set that can be labeled by a countable set is countable.*

Proof. Let $\mathcal{T}$ be a set that is labeled by a countable set $\mathcal{L}$. By definition there is a one-to-one function mapping $\mathcal{T}$ into the set of finite sequences of elements of $\mathcal{L}$, which is a countable set by Corollary 10.3.13. It follows from Corollary 10.3.10 that $\mathcal{T}$ is countable.   □

Any set that is proven to be countable by the Enumeration Principle could, of course, also be proven to be countable without using this principle. However, the Enumeration Principle often leads to very simple proofs.

Theorem 10.3.18. *The set of all rational numbers is countable.*

Proof. As indicated above (Example 10.3.16), the set of rational numbers can be labeled by the set $\mathcal{L} = \{0, 1, 2, 3, 4, 5, 6, 7, 8, 9, -, /\}$. Since $\mathcal{L}$ is finite, the Enumeration Principle (10.3.17) gives the result.   □

You might find the above proof more satisfying than the "zig-zag" proof of the fact that the set of positive rational numbers is countable (Theorem 10.1.13).

Corollary 10.3.19. *The set of all integers is countable.*

Proof. A subset of a countable set is countable (Theorem 10.3.9), so this follows from the previous theorem (10.3.18).   □

Example 10.3.20. The set $\mathcal{T} = \{3 + \sqrt{m} + n^m : m \text{ and } n \text{ are natural numbers}\}$ is countable. To see this, we use the Enumeration Principle (10.3.17). One possible labeling set is $\mathcal{L} = \mathbb{N} \cup \{+, \sqrt{}\}$. Since it is the union of a countable set and a finite set, $\mathcal{L}$ is countable (Thereom 10.2.10). The element $3 + \sqrt{m} + n^m$ of $\mathcal{T}$ can be labeled by the sequence $(3, +, \sqrt{}, m, +, n, m)$, so the Enumeration Principle gives the result.

You may have heard the assertion that π is a "transcendental number"; what does that mean?

Definition 10.3.21. The real number x_0 is said to be *algebraic* if it is the root of a polynomial with integer coefficients. The real number x_0 is said to be *transcendental* if there is no polynomial with integer coefficients that has x_0 as a root. (There are also complex algebraic numbers; see Problem 30 at the end of this chapter.)

For example, the number $-\frac{3}{4}$ is algebraic, since it is a root of the polynomial $4x + 3$. More generally, each rational number $\frac{m}{n}$ is algebraic since it is a root of the polynomial $nx - m$. There are also many irrational numbers that are algebraic, such as $\sqrt{2}$, which is a root of the polynomial $x^2 - 2$, and $\left(\frac{3}{4}\right)^{\frac{1}{5}}$, which is a root of the polynomial $4x^5 - 3$.

It is not so easy to prove the existence of transcendental numbers. It is well known that π is transcendental, but it is very difficult to prove that fact. It is a lot simpler, but still quite difficult, to prove that e, the base of the natural logarithm, is transcendental. It is a very surprising and beautiful fact that it is much easier to prove that most real numbers are transcendental than it is to prove that any specific real number is transcendental. This is a corollary of the following.

Theorem 10.3.22. *The set of real algebraic numbers is countable.*

Proof. We show that the set of real algebraic numbers can be labeled by the integers; the Enumeration Principle (10.3.17) then establishes the theorem. Let x_0 be a real algebraic number. We label x_0 by specifying any polynomial with integer coefficients that has x_0 as a root and then indicating the position that x_0 occupies among the roots of that polynomial. The details of this labeling are as follows.

Let $a_n x^n + a_{n-1} x^{n-1} + \ldots + a_1 x + a_0$ be a polynomial of degree n with integer coefficients that has x_0 as a root. The first terms of the label assigned to x_0 are the coefficients of the polynomial listed according to the descending powers of x. If, for any non-negative integer m less than n, the polynomial does not have a term of degree m, then we list 0 as the corresponding coefficient; that is, $a_m = 0$. The last term in the labeling of x_0 is the natural number k that indicates which position x_0 occupies among all the real roots of the polynomial ordered in the usual way. That is, $k = 1$ if x_0 is the smallest real root of the polynomial, $k = 2$ if x_0 is the second smallest real root, and so on. Thus, the label for x_0 is

$$(a_n, a_{n-1}, \ldots, a_1, a_0, k)$$

In this manner, every real algebraic number is labeled by a finite sequence of integers. Since the set of integers is countable (Corollary 10.3.19), it follows from the Enumeration Principle (10.3.17) that the set of real algebraic numbers is countable. □

The above easily establishes the existence of transcendental numbers.

Corollary 10.3.23. *The set of real transcendental numbers is uncountable.*

Proof. The set of real algebraic numbers is countable (Theorem 10.3.22). If the set of real transcendental numbers was countable, then the set of all real numbers would be the union of two countable sets, and therefore countable (Theorem 10.2.10). Since the set of all real numbers is uncountable (Theorem 10.3.8 and Theorem 10.2.3), the set of transcendental numbers is uncountable.

The cardinality of a finite set consisting of n elements is said to be n. We now introduce some standard notation for the sizes of some of the most common infinite sets.

Definition 10.3.24. We say that the set S *has cardinality* $\aleph_0$ (which we read "aleph naught") if the cardinality of S is the same as that of the natural numbers, in which case we write $|S| = \aleph_0$.

For example, $|\mathbb{Q}| = \aleph_0$.
There is also a standard notation for the cardinality of the set of real numbers.

Definition 10.3.25. We say that the set S *has cardinality* c if the cardinality of S is the same as the cardinality of the set of real numbers; c is sometimes said to be the *cardinality of the continuum*.

For example, $|[3, 9]| = c$.
Note that $\aleph_0 < c$, in the sense that every set with cardinality $\aleph_0$ has cardinality less than every set with cardinality c.

It is important to note that $\aleph_0$ is the smallest infinite cardinality, in the following sense.

Theorem 10.3.26. *If S is an infinite set, then $\aleph_0 \leq |S|$.*

Proof. To establish this, we must show that S has a subset S_0 of cardinality $\aleph_0$. We proceed as follows. Since S is infinite, it surely contains some element, say s_1. Similarly, $S \setminus \{s_1\}$ (i.e., the set obtained from S by removing s_1) contains some element, say s_2. Similarly, $S \setminus \{s_1, s_2\}$ contains some element s_3. Proceeding in this manner creates an infinite sequence $(s_1, s_2, s_3, \ldots)$ of elements of S. Let $S_0 = \{s_1, s_2, s_3, \ldots\}$. Then clearly $|S_0| = |\mathbb{N}| = \aleph_0$. Since S_0 is a subset of S, it follows that $\aleph_0 \leq |S|$. □

Thus, $\aleph_0$ is the smallest infinite cardinality. Is there a largest cardinality?

Definition 10.3.27. If S is any set, then the set of all subsets of S is called the *power set of S* and is denoted $\mathcal{P}(S)$.

The terminology "power set of S" comes from the following theorem. (This was stated as Problem 5 in Chapter 2.)

Theorem 10.3.28. *If S is a finite set with n elements, then the cardinality of $\mathcal{P}(S)$ is 2^n.*

Proof. First note that this is true for $n = 0$. For the only set with 0 elements is $\emptyset$, the empty set. The empty set has one subset, namely itself. Since $2^0 = 1$, the theorem holds for $n = 0$.

We proceed by mathematical induction. Suppose that every set with k elements has 2^k subsets and let S be a set with $k + 1$ elements. Suppose that s_0 is any element of S and let S_0 be the subset $S \setminus \{s_0\}$ of S obtained by removing s_0. Then S_0 has k elements and, by the inductive hypothesis, $|\mathcal{P}(S_0)| = 2^k$. Suppose that $\mathcal{T}$ is any subset of S_0. Then $\mathcal{T}$ is also a subset of S. The set $\mathcal{T} \cup \{s_0\}$ is a different subset of S. Thus, for each subset $\mathcal{T}$ of S_0, there are two subsets of S, $\mathcal{T}$ and $\mathcal{T} \cup \{s_0\}$. It follows that there are twice as many subsets of S as there are subsets of S_0. That is,

$$|\mathcal{P}(S)| = 2 \cdot |\mathcal{P}(S_0)| = 2 \cdot 2^k = 2^{k+1}$$

The theorem follows by mathematical induction.   □

What is the relationship between $|S|$ and $|\mathcal{P}(S)|$ when S is an infinite set?

Theorem 10.3.29. *For every set S, $|S| < |\mathcal{P}(S)|$.*

Proof. It is easy to see that $|S| \leq |\mathcal{P}(S)|$, for among the subsets of S are the "singleton sets;" i.e., sets of the form $\{s\}$, for each $s \in S$. The collection $\mathcal{P}_0$ of all singleton subsets of S is a subset of $\mathcal{P}(S)$. A one-to-one function f mapping S into $\mathcal{P}(S)$ can be defined by $f(s) = \{s\}$, for all s in S. Thus, $|S| = |\mathcal{P}_0|$, so $|S| \leq |\mathcal{P}(S)|$.

To show that $|S| < |\mathcal{P}(S)|$, we must show that there does not exist any one-to-one function f taking S onto $\mathcal{P}(S)$.

Suppose, then, that f is any function taking S into $\mathcal{P}(S)$. We will show that f cannot be onto; that is, that there is an element of $\mathcal{P}(S)$ (i.e., a subset of S) that is not in the range of f.

For each $s \in S$, $f(s)$ is a subset of S. Define the subset S_0 of S by

$$S_0 = \{s \in S : s \notin f(s)\}$$

That is, the subset S_0 of S is defined to consist of all of those elements s of S that are not in the subset of S that f assigns to s.

The set S_0 is an element of $\mathcal{P}(S)$. We will show that it is not in the range of f. To prove this by contradiction, suppose that there was some $s_0 \in S$ such that $f(s_0) = S_0$. We show that this is impossible by considering the question: is s_0 in S_0? We will see that either answer to this question leads to a contradiction.

Suppose that $s_0 \notin S_0$. The definition of S_0 is that it consists of those elements of S that are not in the subsets they are sent to by f. Thus, if s_0 is not in $f(s_0)$, s_0 is in S_0. In other words, $s_0 \notin S_0$ implies $s_0 \in S_0$, which is a contradiction.

On the other hand, if s_0 is in S_0, then the definition of S_0 implies that s_0 is not in $f(s_0)$. But $f(s_0) = S_0$, so $s_0 \notin S_0$. Thus, $s_0 \in S_0$ implies $s_0 \notin S_0$, which is also a contradiction. If there was an s_0 satisfying $f(s_0) = S_0$, then s_0 would either be in S_0 or not be in S_0. Therefore, there is no s_0 satisfying $f(s_0) = S_0$, and the theorem is proven. $\qquad \square$

Corollary 10.3.30. *If S is any set, then there exists a set $\mathcal{T}$ whose cardinality is greater than that of S.*

Proof. By the previous theorem (10.3.29), $\mathcal{T} = \mathcal{P}(S)$ establishes this corollary. $\quad \square$

In particular, for the set of real numbers $\mathbb{R}$, the cardinality of $\mathcal{P}(\mathbb{R})$, the set of all sets of real numbers, is greater than c. Because of the analogy to the case of finite sets, it is standard to write $|\mathcal{P}(\mathbb{R})| = 2^c$.

Similarly, $2^{\aleph_0}$ denotes the cardinality of $\mathcal{P}(\mathbb{N})$. By the above, $\aleph_0 < 2^{\aleph_0}$. Also, as we have seen, $\aleph_0 < c$. What is the relationship between $2^{\aleph_0}$ and c?

Theorem 10.3.31. *The cardinality of the set of all sets of natural numbers is the same as the cardinality of the set of real numbers. That is, $|\mathcal{P}(\mathbb{N})| = c$, or $2^{\aleph_0} = c$.*

Proof. Since $\big|[0, 1]\big| = |\mathbb{R}|$ (Theorem 10.3.8) and $\big|[0, 1]\big| = \big|[0, 1)\big|$ (Theorem 10.3.7), it suffices to prove that $\big|[0, 1)\big| = |\mathcal{P}(\mathbb{N})|$. We require the fact that numbers in $[0, 1)$ can be represented by infinite decimals; that is, expressions such as $.a_1 a_2 a_3 \ldots$ where each a_i is a digit between 0 and 9 (this is shown in Chapter 13; see Theorem 13.6.3). Some numbers have two such representations. For example, $.26999\ldots = .27000\ldots$ (see Section 13.6 of Chapter 13). In such cases, choose the representation ending in a string of 0's.

To show $\big|[0, 1)\big| \leq |\mathcal{P}(\mathbb{N})|$, define the function f from $[0, 1)$ into $\mathcal{P}(\mathbb{N})$ by letting $f(.a_1 a_2 a_3 \ldots)$ be the subset of $\mathbb{N}$ consisting of all natural numbers of the form $a_k 10^k + 1$. That is, $f(.a_1 a_2 a_3 \ldots) = \{a_k 10^k + 1 : k \in \mathbb{N}\}$. Note that the set corresponding to an infinite decimal contains at most one number between 10^k and $9 \cdot 10^k + 1$ for each natural number k. It contains such a number when a_k is not 0. The set contains 1 if and only if some a_k is 0.

To show that f is one-to-one, suppose that $.a_1 a_2 a_3 \ldots$ is not equal to $.b_1 b_2 b_3 \ldots$. Then, for some k, $a_k \neq b_k$. At most one of a_k and b_k is 0. Without loss of generality, assume that a_k is not 0. Then $a_k 10^k + 1$ is in $f(.a_1 a_2 a_3 \ldots)$. However, the only number in $f(.b_1 b_2 b_3 \ldots)$ that could be between 10^k and $9 \cdot 10^k + 1$ is $b_k 10^k + 1$, which is not equal to $a_k 10^k + 1$. Therefore $f(.a_1 a_2 a_3 \ldots)$ is not equal to $f(.b_1 b_2 b_3 \ldots)$. Thus, f is one-to-one, and it follows that $\big|[0, 1)\big| \leq |\mathcal{P}(\mathbb{N})|$.

We now prove that $|\mathcal{P}(\mathbb{N})| \leq \big|[0, 1)\big|$. Define a function g taking $\mathcal{P}(\mathbb{N})$ to $[0, 1)$ by $g(S) = .c_1 c_2 c_3 \ldots$, where $c_j = 1$ if $j \in S$ and $c_j = 0$ if $j \notin S$. It is clear that g is one-to-one, so $|\mathcal{P}(\mathbb{N})| \leq \big|[0, 1)\big|$.

The Cantor–Bernstein Theorem (10.3.5) completes the proof. $\qquad \square$

Definition 10.3.32. The *unit square in the plane* is the subset of the plane consisting of all points whose x and y coordinates are both between 0 and 1. That is, the unit square is the set

$$\{(x, y) : 0 \le x \le 1, 0 \le y \le 1\}$$

Theorem 10.3.33. *The cardinality of the unit square in the plane is* c.

Proof. Let S denote the unit square. It is clear that $|S| \ge c$, since S contains the subset

$$S_0 = \{(x, 0) : 0 \le x \le 1\}$$

and there is an obvious pairing of S_0 with $[0, 1]$.

To establish the reverse inequality, we will construct a one-to-one function f mapping S into $[0, 1]$. We represent the coordinates of points in the unit square as infinite decimals. In ambiguous cases (i.e., where a representation of a number could end in either a string of 0's or a string of 9's), we choose the representation ending in a string of 9's. We then define the function f by

$$f\big((.a_1a_2a_3\ldots, .b_1b_2b_3\ldots)\big) = .a_1b_1a_2b_2a_3b_3\ldots$$

We claim that f is one-to-one. This follows since $f\big((x, y)\big) = .c_1c_2c_3\ldots$ implies that $x = .c_1c_3c_5\ldots$ and $y = .c_2c_4c_6\ldots$. Thus, $|S| \le \big|[0, 1]\big|$, so the Cantor–Bernstein Theorem (10.3.5) gives $|S| = \big|[0, 1]\big|$.   □

It can be interesting to determine the cardinality of various sets of functions. We present one example below; other examples are given in the problems. The following definition will be useful.

Definition 10.3.34. Let S be a set and S_0 be a subset of S. The *characteristic function* of S_0 as a subset of S is the function f, with domain S, defined by $f(s) = 1$ if $s \in S_0$ and $f(s) = 0$ if $s \notin S_0$.

Note that the range of every characteristic function is contained in the two element set $\{0, 1\}$. Conversely, a function with domain S whose range is contained in $\{0, 1\}$ is a characteristic function of a subset of S; namely, the set of all those $s \in S$ that the function takes to 1.

The following is a very easy, but very useful, fact.

Theorem 10.3.35. *For any set S, the set of all characteristic functions with domain S has the same cardinality as $\mathcal{P}(S)$.*

Proof. As indicated in the definition above of characteristic function, each subset does have a characteristic function. On the other hand, if two characteristic functions are equal as functions, they must be characteristic functions of the same subset (the subset consisting of all elements of the set on which the functions have value 1). Thus, the correspondence between the set of subsets of S and characteristic functions with domain S is one-to-one and onto. □

Theorem 10.3.36. *The cardinality of the set of all functions mapping* $[0, 1]$ *into* $[0, 1]$ *is* 2^c.

Proof. Among the functions are those that take on values contained in the two element set $\{0, 1\}$; i.e., the characteristic functions with domain $[0, 1]$. By the previous theorem (10.3.35), this set of characteristic functions has cardinality 2^c. Thus, the set of all functions mapping $[0, 1]$ into $[0, 1]$ has cardinality at least 2^c.

To prove the reverse inequality, we use the fact that every function is determined by its graph. The graph of a function f from $[0, 1]$ to $[0, 1]$ is $\{(x, f(x)) : x \in [0, 1]\}$, which is a subset of the unit square. Clearly, every function has a graph, and if two functions have the same graphs, then they are the same function. Thus, the set of functions we are considering corresponds to a collection of some of the subsets of the unit square and hence has cardinality at most equal to that of the set of all subsets of the unit square. We have seen (Theorem 10.3.33) that the cardinality of the unit square is c. It follows that the cardinality of the set of *all* subsets of the unit square is 2^c. Therefore, the cardinality of the set of graphs of functions (and thus of the set of functions) is at most 2^c. By the Cantor–Bernstein Theorem (10.3.5), the cardinality of the set of functions is 2^c. □

There are some serious deficiencies in the general approach to set theory that we have been describing. The following illustrates some of the problems.

Cantor's Paradox. Let S denote the set of all sets. Then every subset of S is an element of S, since each subset is a set. That is, $\mathcal{P}(S)$ is a subset of S. Hence, $|\mathcal{P}(S)| \leq |S|$. On the other hand, $|S| < |\mathcal{P}(S)|$ (by Theorem 10.3.29). The Cantor–Bernstein Theorem (10.3.5) proves that this is a contradiction.

What does this contradiction mean? If there is a contradiction, then something is false; but what? The only assumption that we have made is that there *is* a set consisting of the set of all sets. This contradiction shows that there cannot be such a set. To avoid Cantor's Paradox, the definition of set has to be more restrictive.

There is another paradox similar to Cantor's.

Russell's Paradox. Define a set to be *ordinary* if it is not an element of itself. (That is, $S \notin S$.) All of the sets that we have discussed so far, except for the set of all sets, are ordinary sets. Each set is, of course, a *subset* of itself, but that is very different from being a member of itself. (For example, the set of natural numbers is not a natural number.)

Let $\mathcal{T}$ denote the set of all ordinary sets. We now ask the question: is $\mathcal{T}$ an ordinary set? If $\mathcal{T}$ was an ordinary set, then, since $\mathcal{T}$ is the set of all ordinary sets, $\mathcal{T} \in \mathcal{T}$. But then $\mathcal{T}$ would not be an ordinary set, since it would be an element of itself. On the other hand, if $\mathcal{T}$ is not an ordinary set, then $\mathcal{T} \in \mathcal{T}$. But every element of $\mathcal{T}$ is an ordinary set, so it would follow that $\mathcal{T}$ is ordinary. That is, if $\mathcal{T}$ is ordinary, it is not ordinary; if $\mathcal{T}$ is not ordinary, it is ordinary. There cannot be such a set.

Note that the Cantor and Russell paradoxes are related in the following sense: the set of all sets, if it existed, would be a set that is not an ordinary set.

When mathematicians became aware of the Cantor and Russell paradoxes, over a hundred years ago, they were very concerned. Why aren't "the set of all sets" and "the set of all ordinary sets" themselves sets? What other "sets" are not really sets?

The above and related paradoxes do not arise when considering sets that generally arise in doing mathematics. Mathematicians have developed several different "axiomatic set theories" in which the concept of "set" is restricted so that the Cantor and Russell paradoxes do not arise. In these set theories, there are no sets that are elements of themselves. The most popular of the axiomatic set theories is called *Zermelo–Fraenkel Set Theory*. The development of axiomatic set theories is fairly complicated and we will not discuss it here. However, the theorems that we presented in this chapter are also theorems in Zermelo–Fraenkel Set Theory although the formal proofs are slightly different.

The following is a very natural question: is there any set S whose cardinality is greater than $\aleph_0$ and less than c? If there is such a set, there would be a one-to-one function taking S into $\mathbb{R}$. Therefore, if there is any such set, then there is a subset of $\mathbb{R}$ with that property. The question can therefore be reformulated: if S is an uncountable subset of $\mathbb{R}$, must the cardinality of S be c? This appears to be a very concrete question. It can be made even more concrete, as follows: if S is a subset of $\mathbb{R}$ and there is no one-to-one function taking S into $\mathbb{N}$, must there exist a one-to-one function taking S onto $\mathbb{R}$?

The Continuum Hypothesis. There is no set with cardinality strictly between $\aleph_0$ and c.

It is very surprising that it is not known whether the Continuum Hypothesis is true or false. It is even more surprising that it has been proven that the Continuum Hypothesis is an undecidable proposition, in the following sense: it has been established that the Continuum Hypothesis can neither be proven nor disproven within standard set theories, such as Zermelo–Fraenkel Set Theory. Mathematicians disagree about the full implications of this. It is our view that it is possible that someone will prove the Continuum Hypothesis in a way that would convince virtually all mathematicians, in spite of it being undecidable within Zermelo–Fraenkel Set Theory. That is, someone might begin a proof as follows: "Let S be an uncountable subset of $\mathbb{R}$. We construct a one-to-one function f mapping S onto $\mathbb{R}$ by first . . . ." Any such proof would have to use something that was not part of Zermelo–Fraenkel Set Theory, since it has been proven that the Continuum Hypothesis cannot be decided within Zermelo–Fraenkel Set Theory. On the other hand, it is our opinion that it is possible that a proof could be found that would be based on properties of the set of real numbers that most mathematicians would agree are true, in spite of the fact that at least one of them would not be part of Zermelo–Fraenkel Set Theory. However, many mathematicians believe that Zermelo–Fraenkel Set Theory captures all the reasonable properties of the real numbers and therefore conclude that no such proof is possible. We invite you to try to prove that those mathematicians are wrong by proving (or disproving) the Continuum Hypothesis. Your chance of success is extremely low, but you might find it interesting to give it a little thought.

10.4 Problems

Basic Exercises

1. Show that the set of all polynomials with rational coefficients is countable.
2. Suppose that the sets S, T, and U satisfy $S \subset T \subset U$ and that $|S| = |U|$. Show that T has the same cardinality as S.
3. Let A and B be countable sets. Prove that the *Cartesian product* of A and B, defined by $A \times B = \{(a, b) : a \in A, b \in B\}$, is countable.
4. Assume that $|A_1| = |B_1|$ and $|A_2| = |B_2|$. Prove:

   (a) $|A_1 \times A_2| = |B_1 \times B_2|$.
   (b) If A_1 is disjoint from A_2 and B_1 is disjoint from B_2, then $|A_1 \cup A_2| = |B_1 \cup B_2|$.

5. Prove that the half-open intervals $[0, 1)$ and $(0, 1]$ have the same cardinality. (This was stated but not proven in Theorem 10.3.7.)
6. What is the cardinality of the set of all functions from $\mathbb{N}$ to $\{1, 2\}$?
7. What is the cardinality of the set of all numbers in the interval $[0, 1]$ that have decimal expansions with a finite number of nonzero digits?
8. Let $\mathbb{Q}(\sqrt{2})$ be the set of real numbers of the form $a + b\sqrt{2}$, where a and b are rational numbers. Find the cardinality of $\mathbb{Q}(\sqrt{2})$.

Interesting Problems

9. Suppose that S and T each have cardinality c. Show that $S \cup T$ also has cardinality c.
10. What is the cardinality of $\mathbb{R}^2 = \{(x, y) : x, y \in \mathbb{R}\}$ (the *Euclidean plane*)?
11. What is the cardinality of the set of all complex numbers?
12. Prove that the set of all finite subsets of $\mathbb{Q}$ is countable.
13. Let S and T be finite sets and let $C = \{f : S \to T\}$ be the set of all functions from S to T. Show that if $|T| > 1$, then $|C| \geq 2^{|S|}$.
14. What is the cardinality of the unit cube, where the unit cube is $\{(x, y, z) : x, y, z \in [0, 1]\}$?
15. What is the cardinality of $\mathbb{R}^3 = \{(x, y, z) : x, y, z \in \mathbb{R}\}$?
16. What is the cardinality of the set of all functions from $\{1, 2\}$ to $\mathbb{N}$?

17. Find the cardinality of the set of all points in $\mathbb{R}^3$ all of whose coordinates are rational.

18. Let S be the set of all functions mapping the set $\left\{\sqrt{2}, \sqrt{3}, \sqrt{5}, \sqrt{7}\right\}$ into $\mathbb{Q}$. What is the cardinality of S?

19. Find the cardinality of the set $\{(x, y) : x \in \mathbb{R},\ y \in \mathbb{Q}\}$.

20. What is the cardinality of the set of all numbers in the interval $[0, 1]$ that have decimal expansions that end with an infinite sequence of 7's?

21. Let t be a transcendental number. Prove that $t^4 + 7t + 2$ is also transcendental.

22. Suppose that $\mathcal{T}$ is an infinite set and S is a countable set. Show that $S \cup \mathcal{T}$ has the same cardinality as $\mathcal{T}$.

23. Let S be the set of real numbers t such that $\cos t$ is algebraic. Prove that S is countably infinite.

24. Let a, b, and c be distinct real numbers. Find the cardinality of the set of all functions mapping $\{a, b, c\}$ into the set of real numbers.

25. What is the cardinality of

$$\left\{ n^{\frac{1}{k}} : n, k \in \mathbb{N} \right\}$$

i.e., the set of all roots of all natural numbers?

26. Prove that there does not exist a set with a countably infinite power set.

27. (This problem requires some basic facts about trigonometric functions.) Find a one-to-one function mapping the interval $(-\frac{\pi}{2}, \frac{\pi}{2})$ onto $\mathbb{R}$.

Challenging Problems

28. (a) Prove directly that the cardinality of the closed interval $[0, 1]$ is equal to the cardinality of the open interval $(0, 1)$ by constructing a function from $[0, 1]$ to $(0, 1)$ that is one-to-one and onto.

    (b) More generally, show that if S is an infinite set and $\{a, b\} \subset S$, then $|S| = |S \setminus \{a, b\}|$. (The notation $S \setminus \{a, b\}$ is used to denote the set of all s in S such that s is not in $\{a, b\}$.)

    [Hint: Use the fact that S has a countably infinite subset containing a and b.]

29. Prove that a set is infinite if and only if it has the same cardinality as a proper subset of itself. (A *proper subset* is a subset other than the set itself.)

30. A complex number is said to be *algebraic* if it is a root of a polynomial with integer coefficients. Prove that the set of all complex algebraic numbers is countable.

31. What is the cardinality of the set of all finite subsets of $\mathbb{R}$?

32. What is the cardinality of the set of all countable sets of real numbers?

33. Find the cardinality of the set of all lines in the plane.

34. Show that the set of all functions mapping $\mathbb{R} \times \mathbb{R}$ into $\mathbb{Q}$ has cardinality 2^c.

35. Prove the following: If x_0 is a real number and n is the smallest natural number such that a polynomial of degree n with integer coefficients has x_0 as a root, and if p and q are polynomials of degree n with integer coefficients that have the same leading coefficients (i.e., coefficients of x^n) and each have x_0 as a root, then $p = q$.

36. Let S be the set of all real numbers that have a decimal representation using only the digits 2 and 6. Show that the cardinality of S is c.

37. Let S denote the collection of all circles in the plane. Is the cardinality of S equal to c or 2^c?

38. Prove that if S is uncountable and $\mathcal{T}$ is a countable subset of S, then the cardinality of $S \setminus \mathcal{T}$ (where $S \setminus \mathcal{T}$ denotes the set of all elements of S that are not in $\mathcal{T}$) is the same as the cardinality of S.

39. Find the cardinality of the set of all polynomials with real coefficients. That is, find the cardinality of the set of all expressions of the form

$$a_n x^n + a_{n-1} x^{n-1} + \cdots + a_1 x + a_0$$

where n is a nonnegative integer (that depends on the expression) and a_0, $a_1, \ldots, a_n$ are real numbers.

40. Prove that the union of c sets that each have cardinality c has cardinality c.

41. Prove that the set of all sequences of real numbers has cardinality c.

Chapter 11
Fundamentals of Euclidean Plane Geometry

In this chapter we describe the fundamentals of Euclidean geometry of the plane. Our approach relies, to some extent, on some intuitively obvious properties of geometric figures that are apparent from looking at diagrams. More rigorous axiomatic treatments of Euclidean geometry are possible.

11.1 Triangles

We begin with a few basic concepts. By a *line* in the plane, we mean a straight line that extends infinitely in two directions; by a *line segment*, we mean the finite part of a line between two given points. We assume as an axiom that, given any two points in the plane, there exists a unique line passing through the two points.

Another basic concept is that of a *triangle*, by which we mean a geometric figure consisting of three points (called its *vertices*) that do not all lie on one line and of the line segments joining those points (which are called the *sides* of the triangle). A typical triangle is pictured in Figure 11.1, where its vertices are labeled with capital letters. We often refer to the sides of the triangle using notation such as AB (or BA), BC, and AC.

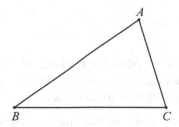

Fig. 11.1 A typical triangle

© Springer Nature Switzerland AG 2018
D. Rosenthal et al., *A Readable Introduction to Real Mathematics*,
Undergraduate Texts in Mathematics, https://doi.org/10.1007/978-3-030-00632-7_11

The triangle in Figure 11.1 might be denoted $\triangle ABC$. The angle at the top of the triangle might be denoted either $\angle A$ or $\angle BAC$. When we say that two line segments are equal we mean that they have the same length. When we say that two angles are equal we mean that they have the same measure (i.e., that one could be placed on top of the other so that they coincide).

Definition 11.1.1. Two triangles are *congruent*, denoted $\cong$, if their vertices can be paired so that the corresponding angles and sides are equal to each other. That is, $\triangle ABC \cong \triangle DEF$ if $\angle A = \angle D$, $\angle B = \angle E$, and $\angle C = \angle F$ and $AB = DE$, $BC = EF$, and $AC = DF$.

If two triangles are congruent, then one can be placed on top of the other so that they completely coincide. More generally, two geometric figures are said to be *congruent* to each other if they can be so placed. It is important to note that congruence of triangles can be established without verifying that all of the pairs of corresponding angles and all the pairs of corresponding sides are equal to each other; as we shall see, equality of certain collections of those pairs implies equality of all of them.

For example, suppose that we fix an angle of a triangle and the lengths of the two sides that form the angle. That is, for example, suppose that, in Figure 11.2, we fix the angle B and lengths AB and BC. It seems intuitively clear that any two triangles with the specified sides AB and BC and the angle B between them are congruent to each other; the only way to complete the given data to form a triangle is by joining A to C by a line segment. Thus, it appears that any two triangles that have two pairs of equal sides and have equal angles formed by those sides are congruent to each other. We assume this as an axiom.

Fig. 11.2 Illustrating side-angle-side

The Congruence Axiom 11.1.2 (Side-Angle-Side). If two triangles have two pairs of corresponding sides equal and also have equal angles between those two sides, then the triangles are congruent to each other.

We speak of this axiom as stating that triangles are congruent if they have "side-angle-side" in common.

Definition 11.1.3. A triangle is said to be *isosceles* if two of its sides have the same length. The angles opposite the equal sides of an isosceles triangle are called the *base angles* of the triangle.

Theorem 11.1.4. *The base angles of an isosceles triangle are equal.*

Proof. Let the given triangle be $\triangle ABC$ with $AB = AC$. Turn the triangle over and denote the corresponding triangle as $\triangle A'C'B'$, as shown in Figure 11.3. Then $\triangle ABC \cong \triangle A'C'B'$. To see this, note that $\angle A = \angle A'$ and $AB = A'C' = AC = A'B'$. Thus, the triangles have side-angle-side in common, and are therefore congruent to each other (11.1.2). In this congruence, $\angle B$ corresponds to $\angle C'$, so $\angle B = \angle C'$. On the other hand, $\angle C'$ was obtained by turning $\angle C$ over, and so $\angle C' = \angle C$. It follows that $\angle B = \angle C$, as was to be proven. $\qquad\square$

Fig. 11.3 Proving that the base angles of an isosceles triangle are equal

Definition 11.1.5. A triangle is *equilateral* if all three of its sides have the same length.

Corollary 11.1.6. *All three angles of an equilateral triangle are equal to each other.*

Proof. Any two angles of an equilateral triangle are the base angles of an isosceles triangle and are therefore equal to each other by the previous theorem. It follows that all three angles are equal. $\qquad\square$

It is sometimes convenient to establish congruence of triangles by correspondences other than side-angle-side.

Theorem 11.1.7 (Angle-Side-Angle). *If two triangles have "angle-side-angle" in common, then they are congruent.*

Proof. Suppose that triangles ABC and DEF are given with $\angle A = \angle D$, $AB = DE$, and $\angle B = \angle E$. If also $AC = DF$, then the triangles are congruent by side-angle-side (11.1.2). We show that this is the case. If AC is not equal to DF, then one of them is shorter; suppose, without loss of generality, that AC is shorter than DF. We will show that is impossible.

Fig. 11.4 Proving angle-side-angle

Mark the length DF along AC beginning with the point A and ending at a point P, as shown in Figure 11.4. Then draw the line connecting B to P. It would follow that $\triangle ABP$ has side-angle-side in common with $\triangle DEF$, so those triangles would be congruent (11.1.2). This would imply that $\angle ABP = \angle E$. But the hypothesis includes the fact that $\angle ABC = \angle E$. This would give $\angle ABC = \angle ABP$, from which we conclude that $\angle PBC = 0$. Therefore, PB lies on BC and hence $AP = AC$. Thus, $AC = DF$ and the theorem is established.   □

If two triangles have equal sides, then they automatically also have equal angles.

Theorem 11.1.8 (Side-Side-Side). *If two triangles have corresponding sides equal to each other, then they are congruent.*

Proof. Let triangles ABC and DEF be given with $AB = DE$, $BC = EF$, and $AC = DF$. At least one of the sides is greater than or equal to each of the other two; suppose, for example, that AB is greater than or equal to each of AC and CB (the other cases would be proven in exactly the same way). Then place the triangle DEF under $\triangle ABC$ so that DE coincides with AB as in Figure 11.5. Connect the points C and F by a straight line. Since $AC = DF$, triangle AFC is isosceles and the base angles ACF and AFC are equal to each other (by Theorem 11.1.4). Similarly, $\triangle BCF$ is isosceles, so $\angle BCF = \angle BFC$. Adding the angles shows that $\angle ACB = \angle DFE$. It follows that triangles ABC and DEF agree in side-angle-side and are therefore congruent to each other (11.1.2).   □

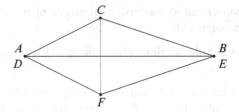

Fig. 11.5 Proving side-side-side

Definition 11.1.9. A *straight angle* is an angle that is a straight line. That is, the angle ABC is a straight angle with vertex B if the points A, B, and C all lie on a straight line and B is in between A and C. A *right angle* is an angle that is half the size of a straight angle.

Definition 11.1.10. *Vertical angles* are pairs of angles that occur opposite each other when two lines intersect. In Figure 11.6, the angles BEA and CED are a pair of vertical angles, and the angles BED and CEA are another pair of vertical angles.

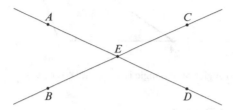

Fig. 11.6 Illustrating vertical angles

Theorem 11.1.11. *Vertical angles are equal.*

Proof. In Figure 11.6, we show that $\angle BEA = \angle CED$, as follows. Angle BEA and angle AEC add up to a straight angle. Angle AEC and angle CED also add up to a straight angle. Therefore, $\angle BEA + \angle AEC = \angle AEC + \angle CED$. Hence, angle BEA equals angle CED. □

One customary way of denoting the size of angles is in terms of degrees.

Definition 11.1.12. The measure of an angle in *degrees* is defined so that a straight angle is $180°$ and other angles are the number of degrees determined by the proportion that they are of straight angles.

In particular, a right angle is $90°$. More generally, if x is the proportion that an angle is of a straight angle, then the degree measure of the angle is given by $180x$.

We will prove that the sum of the angles of a triangle is a straight angle. In the approach that we follow, the following partial result is essential.

Theorem 11.1.13. *The sum of any two angles of a triangle is less than $180°$.*

Proof. Consider an arbitrary triangle ABC as depicted in Figure 11.7 (on the next page) and extend the side AB beyond A to a point F, as shown. We will prove that the sum of angles CAB and ACB is less than a straight angle.

Let M be the midpoint of the side AC. Draw the line from B through M and extend it to the other side of M to a point D such that $DM = MB$. Draw the line from D to A. Then $\angle DMA = \angle CMB$, since they are a pair of vertical angles (Theorem 11.1.11). By construction, $AM = MC$ and $DM = MB$. Thus, $\triangle CMB \cong \triangle AMD$ by side-angle-side (11.1.2). It follows that $\angle DAM$ is equal to $\angle BCM$. Therefore, the sum that we are interested in, $\angle BCM + \angle MAB$, is equal to the sum of $\angle DAM + \angle MAB$. But this latter sum is less than a straight angle, since it together with $\angle DAF$ sums to a straight angle. □

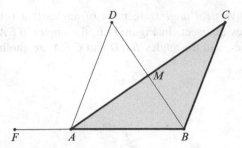

Fig. 11.7 The sum of two angles of a triangle is less than 180 degrees

11.2   The Parallel Postulate

Definition 11.2.1. Two lines in the plane are *parallel* if they do not intersect.

For hundreds of years, mathematicians tried to prove the "Parallel Postulate" as a theorem that followed from the other basic assumptions about Euclidean geometry. Finally, in the 1800s, this was shown to be impossible when different geometries were constructed that satisfied the other basic assumptions but not the following (such geometries are now called "non-Euclidean geometries"). Since it cannot be proven, we assume it as an axiom.

The Parallel Postulate 11.2.2. Given a line and a point that is not on the line, there is one and only one line through the given point that is parallel to the given line.

We will develop necessary and sufficient conditions for two lines to be parallel. Given two lines, a third line that intersects both of the first two is said to be a *transversal* of the two lines. Given a transversal of two lines, a pair of angles created by the intersections of the transversal with the lines are said to be *corresponding angles* if they lie on the same sides of the given lines. In Figure 11.8, T is a transversal of the lines L_1 and L_2. The angles b and d are a pair of corresponding angles. The four angles between the parallel lines are called *interior angles*. If two interior angles lie on opposite sides of the transversal, they are called *alternate interior angles*. In Figure 11.8, the angles b and e are a pair of alternate interior angles, as are the angles c and f.

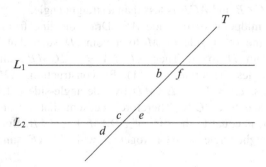

Fig. 11.8 Corresponding angles and alternate interior angles

Theorem 11.2.3. *If the angles in a pair of corresponding angles created by a transversal of two lines are equal to each other, then the two lines are parallel to each other.*

Proof. If the theorem was not true, then there would be a situation as depicted in Figure 11.9, where $\angle a = \angle c$ and lines L_1 and L_2 intersect in some point P. Now $\angle a + \angle b$ is clearly a straight angle. Then, since $\angle a = \angle c$, it would follow that that the sum of angles b and c is a straight angle, contradicting Theorem 11.1.13. Hence, the lines L_1 and L_2 cannot intersect.   □

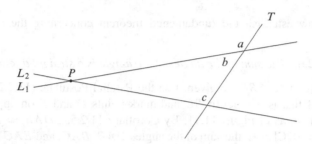

Fig. 11.9 Equal corresponding angles imply that lines are parallel

The converse of this theorem is also true.

Theorem 11.2.4. *If two lines are parallel, then any pair of corresponding angles are equal to each other.*

Proof. Suppose that two lines are parallel and that two corresponding angles differ from each other. Then there would be a situation, such as that depicted in Figure 11.8, with two parallel lines L_1 and L_2 and angle b different from angle d.

Fig. 11.10 If lines are parallel, corresponding angles are equal

Suppose that angle b is bigger than angle d (the proof where this inequality is reversed would be virtually identical). Then, as depicted above in Figure 11.10, we could draw a line L through Q, the point of intersection of L_1 and T, such that angle b' is equal to angle d. But then, by the previous theorem (11.2.3), L

would be parallel to L_2. Thus, L and L_1 would be distinct lines through the point Q which are both parallel to L_2, contradicting the uniqueness aspect of the Parallel Postulate (11.2.2).   □

Corollary 11.2.5. *If two lines are parallel, then any pair of alternate interior angles are equal to each other.*

Proof. Consider parallel lines L_1 and L_2 and alternate interior angles b and e as pictured in Figure 11.8. From Theorem 11.2.4, we know that angles b and d are equal, and, by Theorem 11.1.11, angle d is equal to angle e. Therefore, angles b and e are equal.   □

We can now establish the fundamental theorem concerning the angles of a triangle.

Theorem 11.2.6. *The sum of the angles of a triangle is a straight angle.*

Proof. Let a triangle ABC be given. Use the Parallel Postulate (11.2.2) to pass a line through A that is parallel to BC and mark points D and E on opposite sides of A on that line, as in Figure 11.11. By Corollary 11.2.5, $\angle DAB = \angle ABC$ and $\angle EAC = \angle ACB$. Clearly, the sum of the angles DAB, BAC, and EAC is a straight angle. Therefore, the sum of the angles ABC, BAC, and ACB is also a straight angle.   □

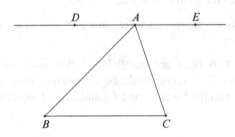

Fig. 11.11   The sum of the angles of a triangle is a straight angle

The following is an obvious corollary.

Corollary 11.2.7. *If two angles of one triangle are respectively equal to two angles of another triangle, then the third angles of the triangles are also equal.*

Corollary 11.2.8 (Angle-Angle-Side). *If two triangles agree in angle-angle-side, then they are congruent.*

Proof. By the previous corollary (11.2.7), the triangles have their third angles equal as well. Thus, the triangles also agree in angle-side-angle and, by Theorem 11.1.7, they are congruent.   □

We finish this section with a useful fact that we will need later in the chapter.

Definition 11.2.9. Lines, or line segments, are said to be *perpendicular* (or *orthogonal*) if they intersect in a right angle.

Lemma 11.2.10. *If two lines are parallel and two other lines are perpendicular to the parallel lines, then the lengths of the perpendicular line segments between the parallel lines are equal to each other.*

Proof. In Figure 11.12, we are assuming that L_1 is parallel to L_2 and that AB and CD are perpendicular to both of L_1 and L_2. (By Theorem 11.2.4, if a line is perpendicular to one of two parallel lines, it is perpendicular to the other as well.) We must prove that $AB = CD$. Note that $\angle ACB = \angle DBC$ and $\angle ABC = \angle BCD$, by Corollary 11.2.5. Thus, the triangles ABC and BCD are congruent by angle-side-angle (11.1.7), since they also share the side BC. Therefore, the corresponding sides AB and CD are equal to each other.   □

Fig. 11.12 Perpendiculars between parallel lines are equal

11.3   Areas and Similarity

We require knowledge of the areas of some common geometric figures.

Recall that a *rectangle* is a four-sided figure in the plane all of whose angles are right angles. By Theorem 11.2.3, this means that the opposite sides of a rectangle are parallel. Lemma 11.2.10 then implies that the opposite sides of a rectangle must have the same length.

The following definition forms the basis for the definition of areas of all geometric figures.

Definition 11.3.1. The *area of a rectangle* is defined to be the product of the lengths of two of its adjacent sides.

The areas of other geometric figures can be obtained either by directly comparing them to rectangles or by approximating them by rectangles.

Definition 11.3.2. A *right triangle* is a triangle one of whose angles is a right angle. The side opposite the right angle in a right triangle is called the *hypotenuse* of the triangle, and the other two sides are called the *legs*.

Theorem 11.3.3. *The area of a right triangle is one-half the product of the lengths of the legs of the triangle.*

Proof. Let the right triangle $\triangle ABC$ be as pictured in Figure 11.13, where $\angle C$ is a right angle. By creating perpendiculars to AC at A and to BC at B, complete the triangle to a rectangle as shown. We will prove that triangles ACB and BDA are congruent and thus have equal areas, which must be half of the area of the rectangle.

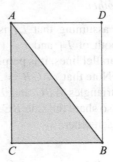

Fig. 11.13 Area of a right triangle

Since the sum of the angles of a triangle is 180° (Theorem 11.2.6), the sum of angles BAC and ABC is 90°. Since AD is perpendicular to AC, the sum of the angles BAC and BAD is also 90°. Hence, $\angle ABC = \angle BAD$. Similarly, since BD is perpendicular to BC, $\angle CAB = \angle ABD$. It follows that $\triangle ABC \cong \triangle BAD$, since they agree in angle-side-angle (11.1.7). Thus, those triangles have equal areas. Since their areas sum to the area of the rectangle whose area is the product of the legs of the triangle ABC, it follows that the area of the triangle ABC is one-half of that product.   □

Definition 11.3.4. Any one of the sides of a triangle may be regarded as a *base* of the triangle. If a side of a triangle is designated as its base, then the *height* of the triangle (relative to that base) is the length of the perpendicular from the base to the vertex of the triangle not on the base. It may be necessary to extend the base of the triangle in order to determine its height, as in the second triangle pictured in Figure 11.14. (In both of the triangles depicted in Figure 11.14, h is the height of the triangle to the base AC.)

Theorem 11.3.5. *The area of any triangle is one-half the product of the length of a base of the triangle and the height of the triangle to that base.*

   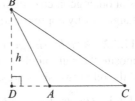

Fig. 11.14 Heights of triangles

Proof. Suppose that the triangle ABC is as pictured in the first triangle in Figure 11.14, where h is the height to the base AC. Then, by the previous theorem (11.3.3), the area of the right triangle ABD is one-half the product of h and AD, and the area of the right triangle DBC is one-half the product of h and DC. The area of triangle ABC is the sum of those areas and is therefore $\frac{1}{2}h \cdot (AD) + \frac{1}{2}h \cdot (DC) = \frac{1}{2}h \cdot (AD + DC) = \frac{1}{2}h \cdot (AC)$. This finishes the proof in this case.

Suppose that $\triangle ABC$ is as pictured in the second triangle in Figure 11.14. The side AC had to be extended to the point D at the bottom of the height. In this case, the area of $\triangle ABC$ is the difference between the area of the right triangle BDC and the area of the right triangle BDA. Hence, the area is $\frac{1}{2}h \cdot (DC) - \frac{1}{2}h \cdot (DA) = \frac{1}{2}h \cdot (AC)$. □

One of the most famous theorems in mathematics is the Pythagorean Theorem. There are very many known proofs of this theorem. One of the nicest, in our view, is the one presented below.

The Pythagorean Theorem 11.3.6. *For any right triangle, the square of the length of the hypotenuse is equal to the sum of the squares of the lengths of the legs.*

Proof. Let the right triangle have legs of lengths a and b and hypotenuse of length c. Place four copies of the given right triangle inside a square whose sides have length $a + b$, as shown in Figure 11.15. We need to prove that the four-sided figure $DEFG$ is a square; i.e., since each of its sides has length c, we must prove that each of its angles is a right angle.

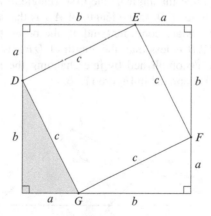

Fig. 11.15 Proof of the Pythagorean Theorem

Since the sum of the angles of a triangle is 180° (Theorem 11.2.6), the sum of the two non-right angles in a right triangle is 90°. Since each angle of $DEFG$ sums with the two non-right angles of the triangle to a straight angle, it follows that each angle of $DEFG$ is 90°. Thus, $DEFG$ is a square, each of whose sides has length c. The area of the big square, each of whose sides has length $a + b$, is the sum of the area of the square $DEFG$ and of the areas of four copies of the original right triangle. That is, $(a + b)^2 = 4(\frac{1}{2}ab) + c^2$. Thus, $a^2 + 2ab + b^2 = 2ab + c^2$, or $a^2 + b^2 = c^2$.   □

Definition 11.3.7. Two triangles are *similar* if their vertices can be paired so that their corresponding angles are equal to each other. We use the notation $\triangle ABC \sim \triangle DEF$ to denote similarity.

Of course (by Corollary 11.2.7), it follows that two triangles are similar if they agree in two of their angles. It is an important, and nontrivial, fact that the corresponding sides of similar triangles are proportional to each other. In other words, if $\triangle ABC \sim \triangle DEF$, then $\frac{AB}{DE} = \frac{AC}{DF} = \frac{BC}{EF}$. The ingenious proof that we present goes back to Euclid. Our basic approach is based on the following lemma.

Lemma 11.3.8. *If a triangle with area S_1 has the same height with respect to a base b_1 that a triangle with area S_2 has with respect to a base b_2, then $\frac{S_1}{b_1} = \frac{S_2}{b_2}$.*

Proof. Let the common height of the two triangles with respect to the given bases be h. Then, $S_1 = \frac{1}{2}hb_1$ and $S_2 = \frac{1}{2}hb_2$. It follows that $\frac{S_1}{b_1} = \frac{1}{2}h = \frac{S_2}{b_2}$.   □

Theorem 11.3.9. *If two triangles are similar, then their corresponding sides are proportional. That is, if $\triangle ABC \sim \triangle DEF$, then $\frac{AB}{DE} = \frac{AC}{DF} = \frac{BC}{EF}$.*

Proof. It suffices to prove that $\frac{AB}{DE} = \frac{AC}{DF}$; the other equation can be obtained as in the proof below but placing the triangles so that the angle at B coincides with the angle at E.

Place the triangles so that the angle of the first triangle at A coincides with the angle of the second triangle at D. If the length of AB is the same as the length of DE, then the two triangles are congruent and all the proportions are 1. Assume, then, that the length of AB is less than the length of DE. (If the opposite is true, the proof below can be accomplished by interchanging the roles of $\triangle ABC$ and $\triangle DEF$.) The situation is depicted in Figure 11.16.

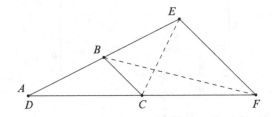

Fig. 11.16 Corresponding sides of similar triangles are proportional

We need to construct triangles to which we can apply the preceding lemma. In Figure 11.16, connect B and F by a line and C and E by a line. Note that, by Theorem 11.2.3, $\angle ABC = \angle DEF$ implies that the line BC is parallel to the line EF. Regard the triangles BEC and BFC as having a common base

BC. Then the corresponding heights of the triangles are the perpendiculars from E to (the extension of) BC and from F to (the extension of) BC, respectively. By Lemma 11.2.10, those heights are equal to each other. Thus, triangles BEC and BFC, having equal bases and heights, have equal areas. Adding those triangles to $\triangle ABC$ establishes that triangles ACE and ABF have equal areas.

We can now use Lemma 11.3.8, as follows. Since $\triangle ABC$ has the same height with respect to its base AB as $\triangle ACE$ has with respect to its base DE, Lemma 11.3.8 implies that

$$\frac{\text{area}(\triangle ABC)}{AB} = \frac{\text{area}(\triangle ACE)}{DE}, \text{ or } \frac{AB}{DE} = \frac{\text{area}(\triangle ABC)}{\text{area}(\triangle ACE)}$$

Similarly, $\triangle ABC$ has the same height with respect to its base AC as $\triangle ABF$ has with respect to its base DF, so

$$\frac{\text{area}(\triangle ABC)}{AC} = \frac{\text{area}(\triangle ABF)}{DF}, \text{ or } \frac{AC}{DF} = \frac{\text{area}(\triangle ABC)}{\text{area}(\triangle ABF)}$$

Since triangles ABF and ACE have the same area, it follows that $\frac{AB}{DE} = \frac{AC}{DF}$. □

Definition 11.3.10. An *angle inscribed in a circle* is an angle whose sides each connect two points of the circle and whose vertex is a point on the circle. (In Figure 11.17, the angle BAC is inscribed in the circle.) The part of the circle that is opposite the angle and lies between the sides of the angle is called the *arc cut off by the angle* (or the *arc intercepted by the angle*).

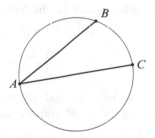

Fig. 11.17 An inscribed angle

We will need the following result in Chapter 12.

Theorem 11.3.11. *If an angle is inscribed in a circle and the arc that it cuts off is a semicircle, then the angle is a right angle.*

Proof. The angles that we are considering are those angles such as $\angle BAC$ in Figure 11.18, where BC is the diameter of the circle and O is the center of the circle. Draw the diameter from A through O.

Since OA and OB are radii of the given circle, and thus are equal, it follows that $\angle OAB = \angle OBA$ (Theorem 11.1.4). Similarly, $\angle OAC = \angle OCA$. Thus, we can label the angles x, y, z, w as indicated in Figure 11.18.

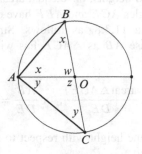

Fig. 11.18   An inscribed angle that cuts off a semicircle is a right angle

Since $2x + w = 180°$ and $z + w = 180°$, it follows that $z = 2x$. Similarly, $w = 2y$. Therefore, $2x + 2y = z + w = 180°$, so $x + y = 90°$. This shows that $\angle BAC$ is $90°$.   □

11.4   Problems

Basic Exercises

1. Which of the following triples cannot be the lengths of the sides of a right triangle?

   (a) $3, 4, 5$ (c) $2, 3, 4$
   (b) $1, 1, 1$ (d) $1, \sqrt{3}, 2$

2. In the diagram given below, lines L_1 and L_2 are parallel and line T is a transversal. If the measure of $\angle d$ is $55°$ and the measure of $\angle f$ is $130°$, find the measures of $\angle b$, $\angle e$, and $\angle c$.

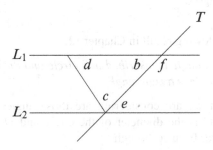

3. In the diagram given below, the line segment BD is perpendicular to the line segment AC, the length of AM is equal to the length of MC, the measure of $\angle C$ is 35°, and the measure of $\angle FAD$ is 111°.

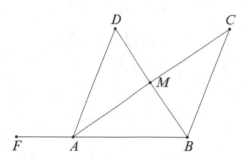

(a) Prove that triangle ABM is congruent to triangle CBM.
(b) Find the measure of $\angle CAB$.
(c) Find the measure of $\angle ABC$.
(d) Find the measure of $\angle ABD$.
(e) Find the measure of $\angle AMD$.
(f) Find the measure of $\angle D$.
(g) Show that the line segments AD and BC are not parallel.

4. Prove that two right triangles are congruent if a leg of one of the triangles has the same length as one of the legs of the other triangle and the lengths of their hypotenuses are equal.

Interesting Problems

5. A *quadrilateral* is a four-sided figure in the plane each of whose interior angles is less than 180°. Prove that the sum of the angles of a quadrilateral is 360°.
6. For quadrilateral $ABCD$, as shown below, suppose that $\angle ABC = \angle CDA$ and $\angle DAB = \angle BCD$. Prove that $AB = CD$ and $BC = AD$.

7. Prove that if two angles of a triangle are equal, then the sides opposite those angles are equal.
8. With reference to the diagram below, prove that $\triangle ABC$ is an isosceles triangle if $\angle DAB = \angle EAC$ and DE is parallel to BC.

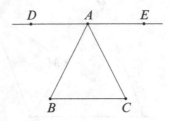

9. A *parallelogram* is a four-sided figure in the plane whose opposite sides are parallel to each other. Prove the following:

   (a) The opposite sides of a parallelogram have the same length.
   (b) The area of a parallelogram is the product of the length of any side and the length of a perpendicular to that side from a vertex not on that side.
   (c) If one of the angles of a parallelogram is a right angle, then the parallelogram is a rectangle.

10. A *trapezoid* is a four-sided figure in the plane two of whose sides are parallel to each other. The *height* of a trapezoid is the length of a perpendicular from one of the parallel sides to the other. Prove that the area of a trapezoid is its height multiplied by the average of the lengths of the two parallel sides.

11. A *square* is a four-sided figure in the plane all of whose sides are equal to each other and all of whose angles are right angles. The *diagonals* of the square are the lines joining opposite vertices. Prove that the diagonals of a square are perpendicular to each other.

12. Show that lines are parallel if there is a transversal such that the alternate interior angles are equal to each other.

Challenging Problems

13. Give an example of two triangles that agree in "angle-side-side" but are not congruent to each other.
14. Find the length of line segment OA in the diagram below (line segment OD is a radius of the circle centered at O, line segment AD has length 4, line segment AC has length 10, and $OABC$ is a rectangle).

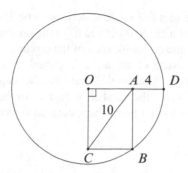

15. Prove the converse of the Pythagorean Theorem; i.e., show that if the lengths of the sides of a triangle satisfy the equation $a^2 + b^2 = c^2$, then the triangle is a right triangle.

16. The following problem establishes some basic results in trigonometry.

    (a) Let θ be any angle between 0 and 90°. Place θ in a right triangle, as shown in the diagram below, and label the sides as in the diagram. Define $\sin \theta$ to be $\frac{a}{c}$, $\cos \theta$ to be $\frac{b}{c}$, and $\tan \theta$ to be $\frac{a}{b}$. Using Theorem 11.3.9, show that these definitions do not depend on which right triangle a given angle θ is placed in.

    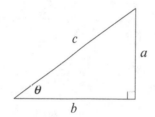

    (b) Label the angles of a triangle with A, B, and C and label the side opposite $\angle A$ with a, the side opposite $\angle B$ with b, and the side opposite $\angle C$ with c. Prove that, in the case where the angles A, B and C are all less than 90 degrees:

    (i) $\dfrac{a}{\sin A} = \dfrac{b}{\sin B} = \dfrac{c}{\sin C}$ (The Law of Sines)

    (ii) $c^2 = a^2 + b^2 - 2ab \cos C$ (The Law of Cosines)

    (The functions sine, cosine and tangent can also be defined for angles greater than 90 degrees. The Law of Sines and the Law of Cosines hold for such angles as well.)

17. (This problem generalizes the result of Theorem 11.3.11.) It is sometimes useful to have a measure of an arc of a circle. One common such measure is in terms of

degrees. The measure of a full circle is defined to be 360 degrees. The number of degrees in any arc of a circle is defined to be the product of 360 and the length of that arc divided by the circumference of the circle.

Prove that the measure of an angle inscribed in a circle is one-half the measure of the arc cut off by the angle. That is, in the diagram below, the number of degrees of $\angle BAC$ is half the number of degrees in the arc BC.

[Hint: One approach is to first prove the special case where AC is a diameter of the circle.]

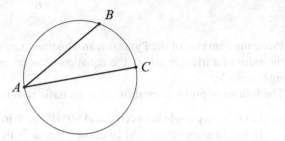

Chapter 12
Constructibility

The Ancient Greeks were interested in many different kinds of mathematical problems. One of the aspects of geometry that they investigated is the question of which geometric figures can be constructed using a compass and a straightedge. A *compass* is an instrument for drawing circles. The compass has two branches that open up like a scissors. One of the branches has a sharp point at the end and the other branch has a pen or pencil at the end. If the compass is opened so that the distance between the two ends is r and the pointed end is placed on a piece of paper and the compass is rotated about that point, the writing end traces out a circle of radius r. The drawing made by any real compass will only approximate a circle of radius r. But we are going to consider constructions theoretically; we will assume that a compass opened up a distance r precisely makes a circle of radius r.

To do geometric constructions, we will also require (as the Ancient Greeks did) another implement. By a *straightedge* we mean a device for drawing lines connecting two points and extending such lines as far as desired in either direction. Sometimes people inaccurately speak of constructions with "ruler and compass." It is important to understand that the constructions investigated by the Ancient Greeks do not allow use of a ruler in the sense of an instrument that has distances marked on it. We can only use such an instrument to connect pairs of points by straight lines; we cannot use it to measure distances.

In this chapter, when we say "construct" or "construction," we always mean "using only a compass and a straightedge."

We begin by showing how to do some basic constructions. But the most interesting part of this chapter will be proving that certain geometric objects cannot be constructed using a straightedge and compass. In particular, we will prove that an angle of 20° cannot be constructed. This implies that an angle of 60° cannot be trisected (i.e., divided into three equal parts) with a straightedge and compass. The Ancient Greeks assumed that there must be some way of trisecting every angle; they thought that they had simply not been clever enough to find a method for doing

© Springer Nature Switzerland AG 2018
D. Rosenthal et al., *A Readable Introduction to Real Mathematics*,
Undergraduate Texts in Mathematics, https://doi.org/10.1007/978-3-030-00632-7_12

so. It was only after mathematical advances in the nineteenth century that it could be proven that there is no way to trisect an angle of 60° with a straightedge and compass. The highlight of this chapter will be a proof of that fact. Although it is hard to imagine how something like that could be proved, we shall see that there is an indirect approach that also establishes many other interesting results.

12.1   Constructions with Straightedge and Compass

Let's start with some very basic constructions.

Definition 12.1.1. A *perpendicular bisector* of a line segment is a line that is perpendicular to the line segment and goes through the middle of the line segment.

Theorem 12.1.2. *Given any line segment, its perpendicular bisector can be constructed.*

Proof. Given a line segment AB, as shown in Figure 12.1, put the point of the compass at A and open the compass to radius the length of AB. Let r equal the length of AB. Then draw the circle with center at A and radius r. Similarly, draw the circle with center at B and radius r. The two circles will intersect at two points; label them C and D as indicated in Figure 12.1. Take the straightedge and draw the line segment from C to D. We claim that CD is a perpendicular bisector of AB.

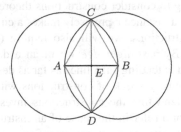

Fig. 12.1 Constructing the perpendicular bisector of a line segment

To prove this, label the point of intersection of CD and AB as E and then draw the line segments AC, CB, BD, and DA. We must prove that $AE = EB$ and that $\angle CEA$ (and/or any of the other three angles at E) is a right angle. First note that AC, CB, BD, and DA all have the same length, r, since they are all radii of the two circles of radius r. Thus, triangle ACD is congruent to triangle BCD, since the third side of each is CD and they therefore agree in side-side-side (11.1.8). It follows that $\angle ACE = \angle BCE$. Hence, triangle ACE is congruent to triangle BCE by side-angle-side (11.1.2). Therefore, $AE = EB$. Moreover, $\angle AEC = \angle BEC$, so, since those two angles sum to a straight angle, each of them is a right angle.   □

Definition 12.1.3. A *bisector of an angle* is a line from its vertex that divides the angle into two equal subangles.

Theorem 12.1.4. *Given any angle, its bisector can be constructed.*

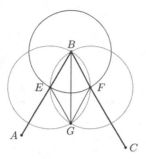

Fig. 12.2 Constructing the bisector of an angle

Proof. Consider an angle ABC, as pictured in Figure 12.2, and draw a circle centered at B that intersects both BA and BC. Label the points of intersection of the circle with AB and with BC as E and F, respectively. Let r be the distance from E to F. Use the compass to draw a circle of radius r centered at E and a circle of radius r centered at F. These two circles intersect in some point G within the angle ABC, as shown in Figure 12.2. Use the straightedge to draw the line segment connecting B to G. We claim that this line segment bisects the angle ABC.

To see this, draw the lines EG and FG. We prove that triangle BEG is congruent to triangle BFG. Note that $BE = BF$, since they are both radii of the original circle centered at B. Note also that $EG = FG$, since they are each radii of circles with radius r. Since triangle BEG and triangle BFG share side BG, it follows from side-side-side (11.1.8) that the two triangles are congruent. Hence, $\angle EBG = \angle FBG$, and BG is a bisector of angle ABC.   □

Theorem 12.1.5. *Any given line segment can be copied using only a straightedge and compass.*

Fig. 12.3 Copying a line segment

Proof. Suppose a line segment AB is given, as pictured in Figure 12.3, and it is desired to copy it on another line. Choose any point C on the other line, and then open the compass to a radius the length of AB. Put the point of the compass at C and draw any portion of the resulting circle that intersects the other line. Label the point of intersection D. Then CD is copy of AB.   □

Theorem 12.1.6. *Any given angle can be copied using only a straightedge and compass.*

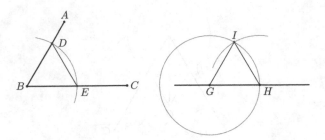

Fig. 12.4 Copying an angle

Proof. Let an angle ABC be given, as in Figure 12.4. We construct an angle equal to $\angle ABC$ with vertex G on any other line. To do this, draw any arc of any circle (of radius, say, r) centered at B that intersects both BA and BC. Label the points of intersection D and E. Draw the circle of radius r centered at G. Use H to label the point where that circle intersects the line containing G. Then adjust the compass to be able to make circles of radius DE. Put the point of the compass at H and draw a portion of the circle that intersects the circle centered at G; call that point of intersection I. Draw line segments connecting D to E and I to H.

Then $IH = DE$, since IH is a radius of a circle with radius DE. Also draw the line segment GI. The lengths of BD, BE, GI, and GH are all equal to r. It follows by side-side-side (11.1.8) that triangle BDE is congruent to triangle GIH. Thus, $\angle IGH$ is a copy of $\angle ABC$.   □

It is sometimes necessary to erect a perpendicular at a given point on a line.

Theorem 12.1.7. *If P is a point on a line, then it is possible to construct a perpendicular to the line that passes through P.*

Proof. Construct a right angle (by, for example, constructing the perpendicular bisector of a line segment, as in Theorem 12.1.2). Then copy the right angle so that its vertex is at P and one side of the angle is the given line. The other side of the angle is then a perpendicular, as desired.   □

Similarly, a perpendicular to a given line can be "dropped" from a point not on the line.

Theorem 12.1.8. *If P is a point that is not on a given line, then it is possible to construct a line through P that is perpendicular to the given line.*

Proof. To drop the perpendicular, make a circle with center at the point P whose radius is large enough that it intersects the line in two points, A and B, as depicted in Figure 12.5. Draw the line segments connecting A to P and B to P. Next, bisect the angle $\angle APB$ (Theorem 12.1.4). The resulting triangles, one on either side of the angle bisector, are congruent to each other since they agree in side-angle-side (11.1.2). This implies that the two angles the angle bisector makes with the original line are equal to each other, and, since they sum to a straight angle, they are therefore each 90 degrees. Hence, the angle bisector is a perpendicular from the point P to the given line.   □

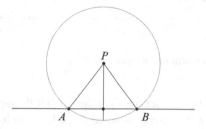

Fig. 12.5 Dropping a perpendicular from a point to a line

Theorem 12.1.9. *If the angles α and β are constructed, then:*

(i) the angle $\alpha + \beta$ can be constructed, and
(ii) for every natural number n, the angle $n\alpha$ can be constructed.

Fig. 12.6 Constructing the sum of two angles

Proof. (i) Let the angles α and β be given, as pictured in Figure 12.6. To construct the angle $\alpha + \beta$, simply copy the angle α with one side DE and the other side above the original angle β, as shown in the third diagram in Figure 12.6.

(ii) This clearly follows from repeated application of part (i), starting with angles α and β that are equal to each other. (This can be proven more formally using mathematical induction.)   □

Theorem 12.1.10. *Given any line segment and any natural number n, the line segment can be divided into n equal parts using only a straightedge and compass.*

Proof. Fix a natural number n. Let a line segment AB be given, as shown in Figure 12.7. Use the straightedge to draw any line segment emanating from A that is at a positive angle less than 90 degrees with AB, and pick a point C on it, as shown. Open the compass to any radius s. Beginning at A, use the compass to mark off n consecutive segments of AC of length s, as illustrated in Figure 12.7. (If the length of AC is less than ns, it is necessary to extend AC to make it greater than ns.) Label the points of intersection of the first $n - 1$ arcs and AC as $P_1, P_2, P_3, \ldots, P_{n-1}$. Label the point of intersection of the line and the nth arc as D. Use a straightedge to

Fig. 12.7 Dividing a line segment into n equal parts

connect D to B. We will construct lines parallel to DB through each point P_j, after which we will show that the intersections of those lines with AB divide AB into n equal segments.

To construct the parallel lines, copy the angle ADB (Theorem 12.1.6) at each point P_j so that one side of the new angle lies on AD and the other side points downwards and is extended to intersect the line AB. These are the dotted lines in Figure 12.7; they are parallel by Theorem 11.2.3. Label the points of intersection of the dotted lines with AB as $Q_1, Q_2, Q_3, \ldots, Q_{n-1}$, as shown.

We claim that the points $Q_1, Q_2, Q_3, \ldots, Q_{n-1}$ divide the segment AB into n equal parts. To see this, note that, for each j, the triangle AP_jQ_j has two angles, $\angle P_jAQ_j$ and $\angle AP_jQ_j$, equal to corresponding angles of $\triangle ADB$. Thus, $\triangle AP_jQ_j$ is similar to $\triangle ADB$ (Corollary 11.2.7). Therefore, the corresponding sides are proportional (Theorem 11.3.9). For each j, the ratio of AP_j to AD is $\frac{j}{n}$. Thus, the length of AQ_j divided by the length of AB is also $\frac{j}{n}$.   □

Therefore, a line segment can be divided into any number of equal parts using only a straightedge and compass. The situation is quite different with respect to angles. In particular, some angles, such as those of 60 degrees, cannot be divided into three equal parts using only a straightedge and compass. We now begin preparation for an indirect approach to establishing that fact.

12.2   Constructible Numbers

We now consider constructing numbers on a number line instead of constructing geometric objects.

We begin by imagining a horizontal line on which a point is arbitrarily marked as 0 and another point, to the right of it, is arbitrarily marked as 1. We consider the question of what other numbers can be obtained by starting with the length 1 (that we take as the distance between the points marked 0 and 1) and doing *geometric constructions* in the plane to obtain other lengths. By a geometric construction, we mean using our straightedge to make lines joining any two points we have already marked (i.e., constructed) or using our compass to construct a circle centered at a constructed point using a radius that has previously been constructed.

Definition 12.2.1. A real number is *constructible* if the point corresponding to it on the number line can be obtained from the marked points 0 and 1 by performing a finite sequence of geometric constructions in the plane using only a straightedge and compass.

Theorem 12.2.2. *Every integer is constructible.*

Proof. The numbers 0 and 1 are given as constructible. The number 2 can easily be constructed: simply take a compass, open it up to radius 1 by placing one side at the point 0 and the other side at the point 1, and then place the pointed side on the point marked 1 and draw the circle of radius 1 with that point as center. The point where that circle meets the number line to the right of 1 is the number 2, so 2 has been constructed. Then clearly 3 can be constructed by placing the compass with radius 1 so as to make a circle centered at 2. Similarly, all the natural numbers can be constructed. To construct the number -1, simply make the circle of radius 1 centered at 0 and mark the intersection to the left of 0 of that circle with the number line. Then -2 can be constructed by marking the point where the circle centered at -1 meets the number line to the left of the point -1. Every negative integer can be constructed similarly.   □

What about the rational numbers?

Theorem 12.2.3. *Every rational number is constructible.*

Proof. To construct, for example, the number $\frac{1}{3}$, simply divide the interval between 0 and 1 into three equal parts (see Theorem 12.1.10) and mark the right-most point of the first part as $\frac{1}{3}$. Similarly, for any natural number n, dividing the unit interval into n equal parts shows that $\frac{1}{n}$ is constructible. Then, for any natural number m, $\frac{m}{n}$ can be constructed by copying m segments of length $\frac{1}{n}$ next to each other on the number line, with the first of those segments beginning at 0.

We have therefore shown that all of the positive rational numbers are constructible. If x is a negative rational number, construct $|x|$ and then make a circle of radius $|x|$ centered at 0; the point to the left of 0 where that circle intersects the number line is x. Thus, every rational number is constructible.   □

We need to get information about the set of all constructible numbers. It is essential to the development of this approach that doing arithmetic with constructible numbers produces constructible numbers.

Theorem 12.2.4. *If a is constructible, then $-a$ is constructible.*

Proof. Place a compass on the number line with its point at 0 and the other end opened to a. Then draw the circle. The number $-a$ will be the point of intersection of the circle and the number line opposite to that of a. (If a is positive, then $-a$ is negative, but if a is negative, then $-a$ is positive.)   □

Theorem 12.2.5. *The sum of two constructible numbers is constructible.*

Proof. Suppose that a and b are constructible. If $b = 0$, then clearly $a+b = a+0 = a$ is constructible. So assume that $b \neq 0$. Open the compass to radius $|b|$, place the point of the compass on the number line at a, and draw the circle. If b is positive, then $a + b$ will be the point of intersection of the circle and the number line to the right of a. If b is negative, then $a + b$ will be the point of intersection of the circle and the number line to the left of a. In both cases, this proves that $a + b$ is constructible.   □

We also need to construct products and quotients. These constructions are a little more complicated; we begin with the following.

Theorem 12.2.6. *If a and b are positive constructible numbers, then $\frac{a}{b}$ is constructible.*

Proof. We consider two possible cases, that where b is greater than 1 and that where b is less than 1.

Fig. 12.8 Constructing quotients (case $b > 1$)

For the case where b is greater than 1, mark the numbers 0, 1, and b on the number line. Use the straightedge to draw a line segment of length greater than a starting from 0, making any angle greater than $0°$ and less than $90°$ with the number line, as pictured in Figure 12.8. Since a is constructible, we can open the compass to radius a. Place the point of the compass at 0 and mark a on the line above the number line. Use the straightedge to connect the point a on the new line to the point b on the number line.

Let β be the angle at b between this line and the number line, and to the left of this line, as shown. Copy the angle β to the point 1 on the number line so that the lower side of the angle is the number line itself. Use the straightedge to extend the other side of the angle beyond the new line. The intersection of the other side of the angle and the new line is a point that we have thereby constructed. Let the distance from the origin to that point be x. We can open the compass to radius x and thereby mark x on the number line. So x is a constructible number. The relationship between x and a and b can be determined by observing that the two triangles formed by the above construction are similar to each other, and therefore the corresponding sides are in proportion (Theorem 11.3.9). It follows that $\frac{x}{a} = \frac{1}{b}$. Thus, $x = \frac{a}{b}$, so we have constructed $\frac{a}{b}$.

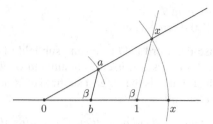

Fig. 12.9 Constructing quotients (case $b < 1$)

The case where b is less than 1 is very similar. In this case, 1 is to the right of b on the number line. Use the straightedge to make a side of an angle starting at 0 above the number line. Since a is constructible, we can open the compass to radius a and mark a point on the new line that is distance a from the vertex of the angle, as in Figure 12.9. Then use the straightedge to draw a straight line between that point and the point b on the number line. Copy the angle, β, at the point b on the number line to the point 1 on the number line and extend the side of the angle so that it intersects the other line. The compass can then be opened to radius equal to the distance from that point of intersection to the origin. If x denotes that radius, then the fact that the corresponding sides of similar triangles are proportional gives $\frac{a}{x} = \frac{b}{1}$, so that $x = \frac{a}{b}$. Thus, $\frac{a}{b}$ is constructible.   □

The above easily leads to the result that products and quotients of constructible numbers are constructible.

Corollary 12.2.7. *If a and b are constructible numbers, then ab is constructible and, if $b \neq 0$, $\frac{a}{b}$ is constructible.*

Proof. First, suppose that a and b are both positive. Then $\frac{a}{b}$ is constructible by the previous theorem (12.2.6). Let $c = \frac{1}{b}$; then c is constructible by the previous theorem using $a = 1$. Since c is constructible, the previous theorem implies that $\frac{a}{c}$ is constructible. But $\frac{a}{c} = \frac{a}{\left(\frac{1}{b}\right)} = ab$, so ab is constructible.

If one or both of a and b is negative, the above can be applied to $|a|$ and $|b|$. Then $ab = |a| \cdot |b|$ if a and b are both negative, and $ab = -|a| \cdot |b|$ if exactly one of them is negative. Similarly, $\frac{a}{b}$ is equal to one of $\frac{|a|}{|b|}$ or $-\frac{|a|}{|b|}$. Since we can construct the negative of any constructible number (Theorem 12.2.4), it follows that ab and $\frac{a}{b}$ are constructible in this case as well.   □

A "field" is an abstract mathematical concept. In this book we do not need to consider general fields; we only need to consider subfields of $\mathbb{R}$. The following definition forms the basis for the rest of this chapter.

Definition 12.2.8. A *subfield of* $\mathbb{R}$ is a set $\mathcal{F}$ of real numbers satisfying the following properties:

(i) The numbers 0 and 1 are both in $\mathcal{F}$.
(ii) If x and y are in $\mathcal{F}$, then $x + y$ and xy are in $\mathcal{F}$ (i.e., $\mathcal{F}$ is "closed under addition" and "closed under multiplication").

(iii) If x is in $\mathcal{F}$, then $-x$ is in $\mathcal{F}$.

(iv) If x is in $\mathcal{F}$ and $x \neq 0$, then $\frac{1}{x}$ is in $\mathcal{F}$.

In this chapter, we use the word *field* to mean "subfield of $\mathbb{R}$." There are many different subfields of $\mathbb{R}$. Of course, $\mathbb{R}$ itself is a subfield of $\mathbb{R}$. So is the set $\mathbb{Q}$ of rational numbers. It is clear that $\mathbb{R}$ is the biggest subfield of $\mathbb{R}$; it is almost as obvious that $\mathbb{Q}$ is the smallest, in the following sense.

Theorem 12.2.9. *If $\mathcal{F}$ is any subfield of $\mathbb{R}$, then $\mathcal{F}$ contains all rational numbers.*

Proof. To see this, first note that 0 and 1 are in $\mathcal{F}$ by property (i) of a subfield of $\mathbb{R}$. Then property (ii) implies that 2 is in $\mathcal{F}$, and 3 is in $\mathcal{F}$, and so on. That is, $\mathcal{F}$ contains all the natural numbers (this can be formally established by a very easy mathematical induction). Property (iii) then implies that $\mathcal{F}$ contains all integers. By property (iv), $\mathcal{F}$ contains the reciprocals of every integer other than 0, so, by property (ii), $\mathcal{F}$ contains all rational numbers.   □

The following is an important fact.

Theorem 12.2.10. *The set of constructible numbers is a subfield of $\mathbb{R}$.*

Proof. This follows immediately from Theorems 12.2.4 and 12.2.5 and Corollary 12.2.7.   □

One of the fundamental theorems in this chapter (Theorem 12.3.12) will provide an alternative characterization of the field of constructible numbers.

Example 12.2.11. The set $\mathbb{Q}(\sqrt{2})$ defined by

$$\mathbb{Q}(\sqrt{2}) = \left\{ a + b\sqrt{2} : a, b \in \mathbb{Q} \right\}$$

is a subfield of $\mathbb{R}$.

Proof. It is clear that $\mathbb{Q}(\sqrt{2})$ contains 0 (since it equals $0 + 0 \cdot \sqrt{2}$) and 1 (since it equals $1 + 0 \cdot \sqrt{2}$). Moreover,

$$\left(a_1 + b_1\sqrt{2} \right) + \left(a_2 + b_2\sqrt{2} \right) = (a_1 + a_2) + (b_1 + b_2)\sqrt{2}$$

Hence, $\mathbb{Q}(\sqrt{2})$ is closed under addition. Furthermore,

$$\left(a_1 + b_1\sqrt{2} \right)\left(a_2 + b_2\sqrt{2} \right) = (a_1 a_2 + 2b_1 b_2) + (a_1 b_2 + a_2 b_1)\sqrt{2}$$

so $\mathbb{Q}(\sqrt{2})$ is closed under multiplication. Also, $-\left(a + b\sqrt{2} \right) = (-a) + (-b)\sqrt{2}$.

It remains to be shown that $\frac{1}{a+b\sqrt{2}}$ is in $\mathbb{Q}(\sqrt{2})$, whenever a and b are not both 0. But,

$$\frac{1}{a + b\sqrt{2}} = \frac{a - b\sqrt{2}}{\left(a + b\sqrt{2} \right)\left(a - b\sqrt{2} \right)} = \frac{a - b\sqrt{2}}{a^2 - 2b^2} = \frac{a}{a^2 - 2b^2} + \frac{-b}{a^2 - 2b^2}\sqrt{2}$$

which is the sum of a rational number and a number that is the product of a rational number and $\sqrt{2}$ and is therefore in $\mathbb{Q}(\sqrt{2})$. (Of course, the above expression would not make sense if $a^2 - 2b^2 = 0$. However, this cannot be the case, since $a^2 - 2b^2 = 0$ would imply $\left(\frac{a}{b}\right)^2 = 2$, and we know that $\sqrt{2}$ is irrational (Theorem 8.2.6).)   □

The field $\mathbb{Q}(\sqrt{2})$ is the field obtained by starting with the field $\mathbb{Q}$ and "adjoining $\sqrt{2}$" to $\mathbb{Q}$; it is called "the extension of $\mathbb{Q}$ by $\sqrt{2}$." This is a special case of a much more general situation.

Theorem 12.2.12. *Let $\mathcal{F}$ be any subfield of $\mathbb{R}$ and let r be any positive number in $\mathcal{F}$. If $\sqrt{r}$ is not in $\mathcal{F}$, then*

$$\mathcal{F}(\sqrt{r}) = \{a + b\sqrt{r} : a, b \in \mathcal{F}\}$$

is a subfield of $\mathbb{R}$.

Proof. The proof is very similar to the proof given above for the special case of $\mathbb{Q}(\sqrt{2})$ (Example 12.2.11). It is easily seen that 0 and 1 are in $\mathcal{F}(\sqrt{r})$, that $\mathcal{F}(\sqrt{r})$ is closed under addition, and that the negative of every element of $\mathcal{F}(\sqrt{r})$ is also in $\mathcal{F}(\sqrt{r})$. To see that it is closed under multiplication, note that

$$\left(a_1 + b_1\sqrt{r}\right)\left(a_2 + b_2\sqrt{r}\right) = (a_1 a_2 + r b_1 b_2) + (a_1 b_2 + a_2 b_1)\sqrt{r}$$

This is in $\mathcal{F}(\sqrt{r})$ since r is in $\mathcal{F}$ and $\mathcal{F}$ itself is a field.
   Also, for any a and b in $\mathcal{F}$,

$$\frac{1}{a + b\sqrt{r}} = \frac{a - b\sqrt{r}}{\left(a + b\sqrt{r}\right)\left(a - b\sqrt{r}\right)} = \frac{a - b\sqrt{r}}{a^2 - rb^2} = \frac{a}{a^2 - rb^2} + \frac{-b}{a^2 - rb^2}\sqrt{r}$$

Note that $a^2 - rb^2 \neq 0$ unless a and b are both 0, because $\sqrt{r} \notin \mathcal{F}$. (If $a^2 - rb^2 = 0$ and $b \neq 0$, then $\left(\frac{a}{b}\right)^2 = r$, and it would follow that $\sqrt{r} \in \mathcal{F}$.) Hence, $\frac{1}{a + b\sqrt{r}}$ is in $\mathcal{F}(\sqrt{r})$ whenever $a + b\sqrt{r}$ is not 0.   □

Definition 12.2.13. If $\mathcal{F}$ is a subfield of $\mathbb{R}$ and r is a positive number that is in $\mathcal{F}$ such that $\sqrt{r}$ is not in $\mathcal{F}$, then the field

$$\mathcal{F}(\sqrt{r}) = \{a + b\sqrt{r} : a, b \in \mathcal{F}\}$$

is the field *obtained by adjoining $\sqrt{r}$ to $\mathcal{F}$* and is called *the extension of $\mathcal{F}$ by $\sqrt{r}$.*

Example 12.2.14. Since $\sqrt{5}$ is not an element of $\mathbb{Q}(\sqrt{2})$, the extension of $\mathbb{Q}(\sqrt{2})$ by $\sqrt{5}$ is

$$\left\{a + b\sqrt{5} : a, b \in \mathbb{Q}(\sqrt{2})\right\} = \left\{\left(c + d\sqrt{2}\right) + \left(e + f\sqrt{2}\right)\sqrt{5} : c, d, e, f \in \mathbb{Q}\right\}$$

For present purposes, we are interested in adjoining square roots to fields of real numbers because that can be done in a "constructible" way.

Theorem 12.2.15. *If r is a positive constructible number, then $\sqrt{r}$ is constructible.*

Proof. Mark the number $r + 1$ on the number line; label it A as in Figure 12.10. Let $M = \frac{r+1}{2}$; M is constructible. Make a circle with center M and radius M. The circle then goes through the point A and also the point corresponding to 0, which we label O. Use D to denote the point corresponding to r on the number line. Erect a perpendicular to the number line at D (Theorem 12.1.7), and let C be the point above the number line at which that perpendicular intersects the circle.

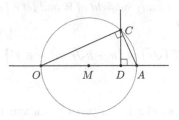

Fig. 12.10 Constructing square roots

The angle OCA is 90°, since it is inscribed in a semicircle (Theorem 11.3.11). Therefore, the sum of the angles OCD and DCA is 90°, from which it follows that the angle COD equals the angle DCA. Thus, triangle OCD is similar to triangle DCA, so their corresponding sides are proportional (Theorem 11.3.9). Let x denote the length of the perpendicular from C to D. Then $\frac{x}{1} = \frac{r}{x}$, so $x^2 = r$. Hence, $x = \sqrt{r}$ and $\sqrt{r}$ is constructible.   □

It follows immediately from this theorem (12.2.15) and the fact that the constructible numbers form a field (Theorem 12.2.10) that every number in $\mathbb{Q}(\sqrt{2})$ is constructible. More generally, every element of $\mathbb{Q}(\sqrt{r})$ is constructible, for every positive rational number r such that $\sqrt{r}$ is irrational. Even more generally, if $\mathcal{F}$ is a field consisting of constructible numbers and r is a positive number in $\mathcal{F}$ such that $\sqrt{r}$ is not in $\mathcal{F}$, then $\mathcal{F}(\sqrt{r})$ consists of constructible numbers. That is, if we start with $\mathbb{Q}$ and keep on adjoining square roots, we get constructible numbers.

Definition 12.2.16. A *tower of fields* is a finite sequence $\mathcal{F}_0, \mathcal{F}_1, \mathcal{F}_2, \ldots, \mathcal{F}_n$ of subfields of $\mathbb{R}$ such that $\mathcal{F}_0 = \mathbb{Q}$ and, for each i from 1 to n, there is a positive number r_i in $\mathcal{F}_{i-1}$ such that $\sqrt{r_i}$ is not in $\mathcal{F}_{i-1}$ and $\mathcal{F}_i = \mathcal{F}_{i-1}(\sqrt{r_i})$.

Note that a tower can be described as a finite sequence $\mathcal{F}_0, \mathcal{F}_1, \mathcal{F}_2, \ldots, \mathcal{F}_n$ of fields of real numbers such that

$$\mathcal{F}_0 \subset \mathcal{F}_1 \subset \mathcal{F}_2 \subset \cdots \subset \mathcal{F}_n$$

with $\mathcal{F}_0 = \mathbb{Q}$ and each $\mathcal{F}_i$ obtained from its predecessor $\mathcal{F}_{i-1}$ by adjoining a square root.

12.3 Surds

We will show that the constructible numbers are exactly those real numbers that are in fields that are in towers. There is a name that is sometimes used for such numbers.

Definition 12.3.1. A *surd* is a number that is in some field that is in a tower. That is, x is a surd if there exists a tower

$$\mathcal{F}_0 \subset \mathcal{F}_1 \subset \mathcal{F}_2 \subset \cdots \subset \mathcal{F}_n$$

such that x is in $\mathcal{F}_n$.

(It should be noted that the word "surd" is sometimes given different meanings. We use the definition given above because we find it most useful.)

Theorem 12.3.2. *The set of all surds is a subfield of* $\mathbb{R}$. *Moreover, if* r *is a positive surd, then* $\sqrt{r}$ *is a surd.*

Proof. To show that the set of surds is a field, it must be shown that the arithmetic operations applied to surds produce surds. This follows immediately if it is shown that, for any surds x and y, there exists a field $\mathcal{F}$ containing both x and y that occurs in some tower. If $\{\sqrt{r_1}, \sqrt{r_2}, \ldots, \sqrt{r_m}\}$ are the numbers adjoined in making a tower that contains x and $\{\sqrt{s_1}, \sqrt{s_2}, \ldots, \sqrt{s_n}\}$ are the numbers adjoined in making a tower containing y, then adjoining all of those numbers produces a field that contains both x and y. Thus, the set of surds is a subfield of $\mathbb{R}$.

To show that square roots of positive surds are surds, let r be a positive surd. Then r is in some field $\mathcal{F}$ that is in a tower. If $\sqrt{r}$ is in $\mathcal{F}$, then $\sqrt{r}$ is clearly a surd. If $\sqrt{r}$ is not in $\mathcal{F}$, then $\sqrt{r}$ is in $\mathcal{F}(\sqrt{r})$, which is in a tower that has one more field than the tower leading to $\mathcal{F}$. ☐

Theorem 12.3.3. *Every surd is constructible.*

Proof. This follows immediately from the theorems that the rational numbers are constructible (Theorem 12.2.3), that the constructible numbers form a field (Theorem 12.2.10), and that the square root of a positive constructible number is constructible (Theorem 12.2.15). ☐

The fundamental theorem that we will need is that the constructible numbers are exactly the surds. Given Theorem 12.3.3, this will follow if it is established that starting with the numbers 0 and 1 and performing constructions with straightedge and compass never produces any numbers that are not surds. Since constructions take place in the plane, we will have to investigate what points in the plane can be constructed.

Definition 12.3.4. We say that the point (x, y) in the plane is *constructible* if that point can be obtained from the points $(0, 0)$ and $(1, 0)$ by performing a finite sequence of constructions with straightedge and compass.

Theorem 12.3.5. *The point* (x, y) *is constructible if and only if both of the coordinates* x *and* y *are constructible numbers.*

Proof. If x and y are constructible numbers, then the point (x, y) can be constructed by constructing the point x on the x-axis, erecting a perpendicular to the x-axis at the point x (Theorem 12.1.7) and constructing y on that perpendicular.

Conversely, if the point (x, y) has been constructed, then the number x can be constructed by dropping a perpendicular from (x, y) to the x-axis (Theorem 12.1.8) and the number y can be constructed by dropping a perpendicular to the y-axis. $\square$

Definition 12.3.6. The *surd plane* is the set of all points (x, y) in the xy-plane such that the coordinates, x and y, are both surds.

By what we have shown above, every point in the surd plane is constructible (Theorems 12.3.3 and 12.3.5). We need to show that every constructible point is in the surd plane.

After we have constructed some points, how can we construct others? We can use a straightedge to make lines joining any two points we have constructed, and we can use a compass to construct a circle centered at a constructible point with a radius that is constructible. New constructible points can then be obtained as points of intersection of lines or circles that we have constructed.

Any one line in the plane has many different equations, as does any one circle. We need to know that there are equations with surd coefficients for all of the lines and circles that arise in constructions.

Theorem 12.3.7. *If a line goes through two points in the surd plane, then there is an equation for that line that has surd coefficients.*

Proof. Suppose that (x_1, y_1) and (x_2, y_2) are distinct points in the surd plane. We consider two cases. If $x_1 \neq x_2$, then

$$y - y_1 = \frac{y_2 - y_1}{x_2 - x_1}(x - x_1)$$

is an equation of the line through the points (x_1, y_1) and (x_2, y_2). Since the surds form a field, the coefficients in this equation are all surds. In the other case, where $x_1 = x_2$, an equation of the line is $x = x_1$.

In both cases, we have shown that an equation for the line through the points (x_1, y_1) and (x_2, y_2) can be expressed in the form $ax + by = c$, where a, b, and c are all surds and a and b are not both 0. $\square$

Theorem 12.3.8. *A circle whose center is in the surd plane and whose radius is a surd has an equation in which the coefficients are all surds.*

Proof. Let the center be (x_1, y_1) and the radius be r. Then one equation of the circle is $(x - x_1)^2 + (y - y_1)^2 = r^2$. Expanding this equation and using the fact that the set of surds is a field shows that this equation has surd coefficients. $\square$

Theorem 12.3.9. *The point of intersection of two distinct nonparallel lines that have equations with surd coefficients is a point in the surd plane.*

Proof. Let such equations be $a_1 x + b_1 y = c_1$ and $a_2 x + b_2 y = c_2$. We consider two cases, that where $a_1 = 0$ and that where $a_1 \neq 0$.

If $a_1 = 0$, then $a_2 \neq 0$ (or else the two lines would be parallel). Then $y = \frac{c_1}{b_1}$, so $a_2 x + b_2 \frac{c_1}{b_1} = c_2$ from which it follows that the intersection of the two lines has coordinates $x = \frac{c_2}{a_2} - \frac{b_2}{a_2} \frac{c_1}{b_1}$ and $y = \frac{c_1}{b_1}$, both of which are surds.

If $a_1 \neq 0$, then $x = -\frac{b_1}{a_1} y + \frac{c_1}{a_1}$. Substituting this in the second equation yields $a_2 \left(-\frac{b_1}{a_1} y + \frac{c_1}{a_1} \right) + b_2 y = c_2$. Since the coefficients are all surds, it is clear that y is also a surd. Hence, so is x, and the theorem is proven in this case as well. $\square$

We next consider the points of intersection of a line and a circle.

Theorem 12.3.10. *The points of intersection of a line that has an equation with surd coefficients and a circle that has an equation with surd coefficients lie in the surd plane.*

Proof. Let $ax + by = c$ and $(x - f)^2 + (y - g)^2 = r^2$ be the equations of a line and a circle, respectively, in which all of the coefficients are surds. Consider first the case where $a = 0$. In this case, $y = \frac{c}{b}$. Substituting this in the equation of the circle yields $(x - f)^2 + (\frac{c}{b} - g)^2 = r^2$. This is a quadratic equation in x. It has 0, 1, or 2 real number solutions depending upon whether the line does not intersect the circle, is tangent to the circle, or intersects the circle in two points. The quadratic formula (Problem 6 in Chapter 9) shows that solutions that exist are obtained from the coefficients by the ordinary arithmetic operations and the extracting of a square root. All of these operations on surds produce surds. Thus any solutions x are surds, proving the theorem in this case.

If $a \neq 0$, then $x = -\frac{b}{a} y + \frac{c}{a}$. Substituting this value in the equation of the circle yields $\left(-\frac{b}{a} y + \frac{c}{a} - f \right)^2 + (y - g)^2 = r^2$. As above, any solutions of this equation are also surds. Therefore the theorem holds in this case too. $\square$

The remaining case is the intersection of two circles.

Theorem 12.3.11. *The points of intersection of two distinct circles that have equations with surd coefficients lie in the surd plane.*

Proof. In order for two distinct circles to intersect, they must have distinct centers. Thus, the equations of the circles can be written in the form $(x - a_1)^2 + (y - b_1)^2 = r_1^2$ and $(x - a_2)^2 + (y - b_2)^2 = r_2^2$, which is equivalent to

$$x^2 - 2a_1 x + a_1^2 + y^2 - 2b_1 y + b_1^2 = r_1^2$$

$$x^2 - 2a_2 x + a_2^2 + y^2 - 2b_2 y + b_2^2 = r_2^2$$

where (a_1, b_1) and (a_2, b_2) are distinct points. (This means that $a_1 \neq a_2$ or $b_1 \neq b_2$.) Subtracting the second equation from the first shows that any point (x, y) that lies on both circles also lies on the line with equation

$$(-2a_1 + 2a_2)x + a_1^2 - a_2^2 + (-2b_1 + 2b_2)y + b_1^2 - b_2^2 = r_1^2 - r_2^2$$

Since this equation has surd coefficients, all points of intersection of this line with either circle lie in the surd plane (Theorem 12.3.10).   □

Theorem 12.3.12. *The field of constructible numbers is the same as the field of surds.*

Proof. We already showed that every surd is constructible (Theorem 12.3.3). On the other hand, Theorems 12.3.7, 12.3.8, 12.3.9, 12.3.10, and 12.3.11 show that starting with surd points in the plane and doing geometric constructions produces only surd points. Thus, every constructible point in the plane has surd coordinates. Since every constructible number is a coordinate of a constructible point in the plane (Theorem 12.3.5), it follows that every constructible number is a surd.   □

This characterization of the constructible numbers is the key to the proof that certain angles cannot be trisected. One of the relationships between constructible angles and constructible numbers can be obtained using the trigonometric function cosine, as follows. We restrict the discussion to acute angles (i.e., angles less than a right angle) simply to avoid having to describe several cases.

Theorem 12.3.13. *The acute angle θ is constructible with a straightedge and compass if and only if $\cos \theta$ is a constructible number.*

Proof. Suppose first that the angle θ is constructible. Copy the angle so that its vertex lies at the point 0 on the number line, one of its sides is the positive part of the number line and the other side is on top of it, as in Figure 12.11. Use the compass to mark a point on the upper side of the angle that is one unit from the point 0. Drop a perpendicular from that point to the number line (Theorem 12.1.8). Then that perpendicular meets the number line at $\cos \theta$, so $\cos \theta$ is constructed.

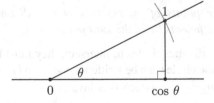

Fig. 12.11 Constructing the cosine of an angle

Conversely, if $\cos \theta$ is constructed, erect a perpendicular upwards from the point $\cos \theta$ on the number line (Theorem 12.1.7). Construct the number $a = \sqrt{1 - \cos^2 \theta}$ and mark the point on the perpendicular with that distance above the number line, as in Figure 12.12. Connecting the point 0 to that marked point by a straightedge produces the angle θ.   □

Fig. 12.12   Constructing an angle from its cosine

With this background and some other preliminary results, we will be able to determine exactly which angles with an integral number of degrees are constructible (see Theorem 12.4.13). First note the following.

Theorem 12.3.14.   *An angle of 60° is constructible.*

Proof. This is an immediate consequence of Theorem 12.3.13, for the cosine of 60° equals $\frac{1}{2}$, and $\frac{1}{2}$ is a constructible number.

There is also an easy direct proof: simply construct an equilateral triangle using a straightedge and compass; each angle of the equilateral triangle is 60°. To construct an equilateral triangle, draw a circle of any radius, call it r, centered at a point A. Draw a line through A that intersects the circle and label a point of intersection B, as in Figure 12.13. Next draw a circle of radius r centered at B and label a point of intersection of the two circles with C. Now draw the segments AC and BC. Triangle ABC is an equilateral triangle (all of whose sides have length r). □

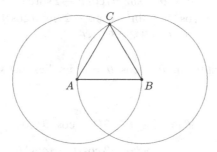

Fig. 12.13   Constructing an equilateral triangle

Corollary 12.3.15.   *The following angles are all constructible: 30°, 15°, 45°, and 75°.*

Proof. We begin with the fact that an angle of 60° is constructible (Theorem 12.3.14). An angle of 30° can be constructed by bisecting an angle of 60°, and an angle of 15° can be constructed by bisecting an angle of 30° (Theorem 12.1.4). An angle of 45° can be constructed by adding an angle of 15° to one of 30°, and an angle of 75° can be constructed by adding an angle of 15° to an angle of 60° (Theorem 12.1.9). □

The material about constructible numbers was developed primarily to prove that some angles are not constructible. We need some additional preliminary results.

Theorem 12.3.16. *For any angle θ, $\cos(3\theta) = 4\cos^3\theta - 3\cos\theta$.*

Proof. Recall the addition formulae for cosine and sine:

$$\cos(\theta_1 + \theta_2) = \cos\theta_1\cos\theta_2 - \sin\theta_1\sin\theta_2$$

and

$$\sin(\theta_1 + \theta_2) = \sin\theta_1\cos\theta_2 + \sin\theta_2\cos\theta_1$$

In particular, if $\theta = \theta_1 = \theta_2$, then

$$\cos(2\theta) = \cos^2\theta - \sin^2\theta$$

and

$$\sin(2\theta) = 2\sin\theta\cos\theta$$

Therefore,

$$\begin{aligned}
\cos(3\theta) &= \cos(2\theta + \theta)\\
&= \cos(2\theta)\cos\theta - \sin(2\theta)\sin\theta\\
&= (\cos^2\theta - \sin^2\theta)\cos\theta - 2\sin\theta\cos\theta\sin\theta\\
&= \cos^3\theta - \sin^2\theta\cos\theta - 2\sin^2\theta\cos\theta\\
&= \cos^3\theta - 3\sin^2\theta\cos\theta
\end{aligned}$$

The trigonometric identity $\sin^2\theta + \cos^2\theta = 1$ implies that $\sin^2\theta = 1 - \cos^2\theta$, which gives

$$\begin{aligned}
\cos(3\theta) &= \cos^3\theta - 3(1 - \cos^2\theta)\cos\theta\\
&= \cos^3\theta - 3\cos\theta + 3\cos^3\theta\\
&= 4\cos^3\theta - 3\cos\theta
\end{aligned}$$

Therefore, $\cos(3\theta) = 4\cos^3\theta - 3\cos\theta$.   □

The case where θ is an angle of $20°$ is of particular interest.

Corollary 12.3.17. *If $x = 2\cos(20°)$, then $x^3 - 3x - 1 = 0$.*

Proof. Using the formula for $\cos(3\theta)$ given above and the fact that the cosine of $60°$ equals $\frac{1}{2}$, we have $\frac{1}{2} = 4\cos^3(20°) - 3\cos(20°)$. This is equivalent to the equation

$$8\cos^3(20°) - 6\cos(20°) - 1 = 0$$

Since $x = 2\cos(20°)$, $x^3 - 3x - 1 = 0$. □

We will show that the cubic equation $x^3 - 3x - 1 = 0$ does not have a constructible root. We need some preliminary results.

Theorem 12.3.18. *If the roots of the cubic equation $x^3 + bx^2 + cx + d = 0$ are r_1, r_2, and r_3, then $b = -(r_1 + r_2 + r_3)$. (It is possible that two or even three of the roots are the same as each other.)*

Proof. By the Factor Theorem (9.3.6), and the fact that the coefficient of x^3 is 1, the cubic equation is the same as $(x - r_1)(x - r_2)(x - r_3) = 0$. Multiplying this out shows that the coefficient of x^2 is $-(r_1 + r_2 + r_3)$; i.e., $b = -(r_1 + r_2 + r_3)$. □

We need the concept of a conjugate for elements of $\mathcal{F}(\sqrt{r})$, analogous to the conjugate of a complex number.

Definition 12.3.19. If $a + b\sqrt{r}$ is an element of $\mathcal{F}(\sqrt{r})$, then the *conjugate* of $a + b\sqrt{r}$, denoted by placing a bar on top of the number, is

$$\overline{a + b\sqrt{r}} = a - b\sqrt{r}$$

Theorem 12.3.20. *The conjugate of the sum of two elements of $\mathcal{F}(\sqrt{r})$ is the sum of the conjugates, and the conjugate of the product of two elements of $\mathcal{F}(\sqrt{r})$ is the product of the conjugates.*

Proof. For the first assertion, simply note that

$$\begin{aligned}
\overline{(a + b\sqrt{r}) + (c + d\sqrt{r})} &= \overline{(a + c) + (b + d)\sqrt{r}} \\
&= (a + c) - (b + d)(\sqrt{r}) \\
&= (a - b\sqrt{r}) + (c - d\sqrt{r}) \\
&= \overline{(a + b\sqrt{r})} + \overline{(c + d\sqrt{r})}
\end{aligned}$$

For products, note that

$$\begin{aligned}
\overline{(a + b\sqrt{r})(c + d\sqrt{r})} &= \overline{(ac + rbd) + (ad + bc)\sqrt{r}} \\
&= (ac + rbd) - (ad + bc)\sqrt{r}
\end{aligned}$$

and

$$\begin{aligned}
\overline{(a + b\sqrt{r})} \cdot \overline{(c + d\sqrt{r})} &= (a - b\sqrt{r})(c - d\sqrt{r}) \\
&= (ac + bdr) - (ad + bc)\sqrt{r}
\end{aligned}$$

Therefore, $\overline{(a + b\sqrt{r})(c + d\sqrt{r})} = \overline{(a + b\sqrt{r})} \cdot \overline{(c + d\sqrt{r})}$. □

Theorem 12.3.21. *If $a + b\sqrt{r}$ is in $\mathcal{F}(\sqrt{r})$ and is a root of a polynomial with rational coefficients, then $a - b\sqrt{r}$ is also a root of the polynomial.*

Proof. Suppose that $a_n(a+b\sqrt{r})^n+a_{n-1}(a+b\sqrt{r})^{n-1}+\cdots+a_1(a+b\sqrt{r})+a_0=0$. Then,

$$\overline{a_n(a+b\sqrt{r})^n+a_{n-1}(a+b\sqrt{r})^{n-1}+\cdots+a_1(a+b\sqrt{r})+a_0} = \overline{0} = 0$$

Since each of the coefficients a_k is rational, $\overline{a_k} = a_k$, for every k. Using this fact and the facts that the conjugate of a sum is the sum of the conjugates and the conjugate of a product is the product of the conjugates (Theorem 12.3.20), it follows that $a_n\overline{(a+b\sqrt{r})}^n+a_{n-1}\overline{(a+b\sqrt{r})}^{n-1}+\cdots+a_1\overline{(a+b\sqrt{r})}+a_0 = 0$. Thus, $a-b\sqrt{r} = \overline{a+b\sqrt{r}}$ is also a root of the polynomial.   □

Theorem 12.3.22. *If a cubic equation with rational coefficients has a constructible root, then the equation has a rational root.*

Proof. Dividing through by the leading coefficient, we can assume that the coefficient of x^3 is 1. Then, by Theorem 12.3.18, the sum of the three roots of the cubic equation is the negative of the coefficient of x^2, and is therefore rational.

We first show that if the equation has a root in any $\mathcal{F}(\sqrt{r})$, then it has a root in $\mathcal{F}$. To see this, suppose the equation has a root in $\mathcal{F}(\sqrt{r})$ of the form $a + b\sqrt{r}$ with $b \neq 0$. Then, by Theorem 12.3.21, the conjugate $a - b\sqrt{r}$ is also a root. If r_3 is the third root and s is the sum of all three roots, then $s = r_3 + (a+b\sqrt{r}) + (a-b\sqrt{r}) = r_3 + 2a$. Thus, $r_3 = s - 2a$. Since $\mathcal{F}$ contains all rational numbers and s is rational, s is in $\mathcal{F}$. Since a is also in $\mathcal{F}$, it follows that the root r_3 is in $\mathcal{F}$ itself.

The preliminary result obtained in the previous paragraph allows us to prove the theorem, as follows. If the polynomial has a constructible root, then, since every constructible number is a surd (Theorem 12.3.12), the root is in a field that occurs at the end of a tower. Consider the field at the end of the shortest tower that contains any root of the given cubic equation. We claim that field is $\mathbb{Q}$. To see this, simply note that if that field was $\mathcal{F}(\sqrt{r})$, the previous paragraph would imply that there is a root in $\mathcal{F}$, which would be at the end of a shorter tower than $\mathcal{F}(\sqrt{r})$ is in. Hence, that field is $\mathbb{Q}$. Thus, the equation has a rational root.   □

We can now prove that an angle of 20° cannot be constructed.

Theorem 12.3.23. *An angle of 20° cannot be constructed with straightedge and compass.*

Proof. If an angle of 20° could be constructed with straightedge and compass, then $\cos(20°)$ would be a constructible number (Theorem 12.3.13). Then $2\cos(20°)$ would also be a constructible number, and the polynomial $x^3 - 3x - 1 = 0$ would therefore have a constructible root (Corollary 12.3.17). It follows from the previous theorem (12.3.22) that this polynomial would have a rational root. Thus, to establish that an angle of 20° is not constructible, all that remains to be shown is that the polynomial $x^3 - 3x - 1 = 0$ does not have a rational root.

Suppose that m and n are integers with $n \neq 0$ and that $\frac{m}{n}$, written in lowest terms, is a root of the equation $x^3 - 3x - 1 = 0$. Then, by the Rational Roots

Theorem (8.2.14), m divides -1 and n divides 1. Thus, m is either 1 or -1 and n is either 1 or -1. Hence, $\frac{m}{n}$ is 1 or -1. Therefore, the only possible rational roots of $x^3 - 3x - 1 = 0$ are $x = 1$ or $x = -1$. Substituting those values for x into the equation shows that neither is a root, so the theorem is proven. $\square$

Corollary 12.3.24. *An angle of 60° cannot be trisected with straightedge and compass.*

Proof. As we have seen, an angle of 60° can be constructed with a straightedge and compass (Theorem 12.3.14). If an angle of 60° could be trisected with straightedge and compass, then an angle of 20° would be constructible. But an angle of 20° is not constructible, by the previous theorem (12.3.23). $\square$

12.4 Constructions of Geometric Figures

Another problem that the Ancients Greeks raised but could not solve was what they called *duplication of the cube*. This was the question of whether or not a side of a cube of volume 2 could be constructed by straightedge and compass.

Theorem 12.4.1. *The side of a cube of volume 2 cannot be constructed with a straightedge and compass.*

Proof. If x is the length of the side of a cube of volume 2, then, of course, $x^3 = 2$, or $x^3 - 2 = 0$. By Theorem 12.3.22, this equation has a constructible root if and only if it has a rational root. Since the cube root of 2 is irrational (Problem 13 in Chapter 8), there is no constructible root. Therefore, the cube cannot be "duplicated" using only a straightedge and compass. $\square$

The question of which regular polygons can be constructed is very interesting.

Definition 12.4.2. A *polygon* is a figure in the plane consisting of line segments that bound a finite portion of the plane. A *regular polygon* is a polygon all of whose angles are equal and all of whose sides are equal.

An equilateral triangle is a regular polygon with three sides. Equilateral triangles can easily be constructed with straightedge and compass (see the proof of Theorem 12.3.14).

A square is a regular polygon with four sides. It is also very easy to construct a square. Simply use the straightedge to draw any line segment, and erect perpendiculars at each end of the line segment (Theorem 12.1.7). Then use the compass to "measure" the length of the line segment and mark points which are that distance above the original line segment on each of the perpendiculars. Using the straightedge to connect those points yields a square.

For each natural number n greater than or equal to 3, there exists a regular polygon with n sides. This can be seen as follows. (Which of these regular polygons is constructible is a more difficult question that we discuss in Theorem 12.4.5.)

Theorem 12.4.3. *For each natural number n greater than or equal to 3, there is a regular polygon with n sides inscribed in a circle.*

Proof. Given a natural number n greater than or equal to 3, take a circle and draw (although it may not be possible to construct) successive adjacent angles of $\frac{360}{n}$ degrees at the center, as shown in Figure 12.14 for the case $n = 6$. Then draw the line segments connecting adjacent points determined by the sides of the angles intersecting the circumference of the circle. We must show that those line segments are all equal in length and that the angles formed by each pair of adjacent line segments are equal to each other.

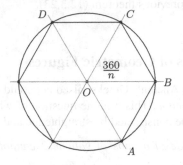

Fig. 12.14 Existence of regular polygons

Consider, for example, the triangles OAB and OCD in Figure 12.14. The angles AOB and COD are each equal to $\frac{360}{n}$ degrees. The sides OA, OB, OC, and OD are all radii of the given circle and are therefore equal to each other. It follows that $\triangle OAB$ is congruent to $\triangle OCD$ by side-angle-side (11.1.2). The same proof shows that all of the triangles constructed are congruent to each other. It follows that all of the sides of the polygon, which are the sides opposite the angles of $\frac{360}{n}$ degrees in the triangles, are equal to each other. The angles of the polygon are angles such as $\angle ABC$ and $\angle BCD$ in the diagram. Each of them is the sum of two base angles of the drawn triangles, and, therefore, the angles of the polygon are equal to each other as well.   □

Definition 12.4.4. *A central angle* of a regular polygon with n sides is an angle of $\frac{360°}{n}$ that has a vertex at the center of the polygon, as in the above proof.

Theorem 12.4.5. *A regular polygon is constructible if and only if its central angle is a constructible angle.*

Proof. Suppose that a regular polygon can be constructed with straightedge and compass. Then its center (a point equidistant from all of its vertices) can be constructed as the point of intersection of the perpendicular bisectors of two adjacent sides of the polygon (see Problem 13 at the end of this chapter). Now the central angle can be constructed as the angle formed by connecting the center to two adjacent vertices of the polygon. All such angles are equal to each other, since the corresponding triangles are congruent by side-side-side (11.1.8). There are n such angles, the sum of which is 360 degrees, so each central angle is $\frac{360}{n}$°.

Conversely, suppose that, for some natural number $n \geq 3$, an angle of $\frac{360}{n}$° is constructible. Then a regular polygon with n sides can be constructed as follows. Construct a circle. Construct an angle of $\frac{360}{n}$° with vertex at the center of the circle. Then construct another such angle adjacent to the first, and so on until n such angles have been constructed. Connecting the adjacent points of intersection of the sides of those angles with the circle constructs a regular polygon with n sides (as shown in the proof of Theorem 12.4.3). $\qquad\square$

Corollary 12.4.6. *A regular polygon with 18 sides cannot be constructed with a straightedge and compass.*

Proof. A regular polygon with 18 sides has a central angle of $\frac{360}{18} = 20$ degrees. We proved in Theorem 12.3.23 that an angle of 20° is not constructible, so the previous theorem implies that a regular polygon with 18 sides is not constructible. $\qquad\square$

Theorem 12.4.7. *If m is a natural number greater than 2, then a regular polygon with 2m sides is constructible if and only if a regular polygon with m sides is constructible.*

Proof. Using Theorem 12.4.5, the result follows by either bisecting or doubling the central angle of the already constructed polygon. $\qquad\square$

Corollary 12.4.8. *A regular polygon with 9 sides is not constructible.*

Proof. This follows immediately from the fact that a regular polygon with 18 sides is not constructible (Corollary 12.4.6) and the above theorem (12.4.7). $\qquad\square$

It is useful to make the following connection between constructible polygons and constructible numbers.

Theorem 12.4.9. *A regular polygon with n sides is constructible if and only if the length of the side of a regular polygon with n sides that is inscribed in a circle of radius 1 is a constructible number.*

Proof. In the first direction suppose that a regular polygon with n sides is constructible. Then such a polygon can be constructed so that it is inscribed in a circle of radius 1 (for example, by putting its constructible central angle in a circle of radius 1). The length of the side can be constructed by using the compass to "measure" the side of the constructed polygon.

Conversely, if s is a constructible number and s is the length of the side of a regular polygon with n sides inscribed in a circle of radius 1, then the regular polygon can be constructed by marking any point on the circle and then using the compass to successively mark points that are at distance s from the previously marked one. The marked points will be vertices of a regular polygon with n sides. $\qquad\square$

Can a pentagon (a regular polygon with 5 sides) be constructed using only a straightedge and compass? The answer is affirmative, but this is not at all easy to see directly. We will approach this by first considering a regular polygon with 10 sides.

Theorem 12.4.10. *A regular polygon with 10 sides is constructible.*

Proof. By Theorem 12.4.9, it suffices to show that the length of a side of such a polygon inscribed in a circle of radius 1 is a constructible number. We determine the length of such a side by using a little geometry. The central angle of a regular polygon with 10 sides is 36°. Consider such an angle with vertex O at the center of a circle of radius 1, as shown in Figure 12.15. Label the points of intersection of the sides of that central angle with the circle A and B. Let s denote the length of the line segment from A to B, and let AC be the bisector of $\angle OAB$. Since $\angle OAB$ is 72° (the sum of the degrees of the equal angles OAB and ABO must be 180° − 36°), it follows that angles OAC and CAB are each 36°. Also, $\angle OBA$ is 72°. Thus, triangles OAB and ABC are similar to each other, so their corresponding sides are in proportion (Theorem 11.3.9). Therefore, triangle ABC is isosceles, and AC has length s.

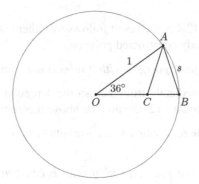

Fig. 12.15 The side of a ten-sided regular polygon

Since $\angle AOB = 36° = \angle OAC$, $\triangle OAC$ is also isosceles. Thus, OC has length s, from which it follows that BC has length $1 - s$. Since the corresponding sides of triangles OAB and ABC are in proportion, it follows that

$$\frac{s}{1-s} = \frac{1}{s}$$

Hence, the length we are interested in, s, satisfies the equation $s^2 = 1 - s$, or $s^2 + s - 1 = 0$. The positive solution of this equation (s is a length) is $\frac{-1+\sqrt{5}}{2}$. Therefore s is a constructible number (Theorem 12.3.12), from which it follows that the regular polygon with 10 sides is constructible.   □

Corollary 12.4.11. *A regular pentagon is constructible.*

Proof. This follows immediately from the above theorem and Theorem 12.4.7. □

Which regular polygons are constructible? As we have shown, those with 3, 4, and 5 sides are, and thus so are those with 6, 8, and 10 sides (Theorem 12.4.7). We proved that a regular polygon with 9 sides is not constructible (Corollary 12.4.8).

What about a polygon with 7 sides? We can approach this question using some facts that we learned about complex numbers. As follows immediately from a previous result (Example 9.2.13), for each natural number n greater than 2, the complex solutions to the equation $z^n = 1$ are the vertices of an n-sided regular polygon inscribed in a circle of radius 1. We will approach the problem by considering the solutions of $z^7 = 1$.

Theorem 12.4.12. *A regular polygon with 7 sides is not constructible.*

Proof. If a regular polygon with 7 sides was constructible, then one could be constructed inscribed in a circle of radius 1 centered at the origin, such that one of the vertices lies on the x-axis at the point corresponding to the number 1. Then the vertices are the 7th roots of unity (Example 9.2.13); that is, they satisfy $z^7 = 1$.

We will analyze the first vertex above the x-axis. Let that vertex lie at the complex number z_0. If the regular polygon was constructible, then z_0 would be a constructible point, and therefore the real part of z_0 would be constructible (simply use Theorem 12.1.8 to construct a perpendicular from z_0 to the x-axis). It would follow that twice the real part is constructible. Let x_0 be twice that real part. We will show that x_0 satisfies a cubic equation that is not satisfied by any constructible number.

Begin by observing that $x_0 = z_0 + \overline{z_0}$. Since $|z_0| = 1$, it follows that $1 = |z_0|^2 = z_0\overline{z_0}$. Thus, $\overline{z_0} = \frac{1}{z_0}$, so $x_0 = z_0 + \frac{1}{z_0}$. The cubic equation satisfied by x_0 will be obtained from the equation $z_0^7 = 1$ and the fact that $z_0 \neq 1$. Note that

$$z_0^7 - 1 = (z_0 - 1)\left(z_0^6 + z_0^5 + z_0^4 + z_0^3 + z_0^2 + z_0 + 1\right)$$

Since $z_0 - 1 \neq 0$,

$$z_0^6 + z_0^5 + z_0^4 + z_0^3 + z_0^2 + z_0 + 1 = 0$$

Dividing through by z_0^3 yields

$$z_0^3 + z_0^2 + z_0 + 1 + \frac{1}{z_0} + \frac{1}{z_0^2} + \frac{1}{z_0^3} = 0$$

Note that $\left(z_0 + \frac{1}{z_0}\right)^3 = z_0^3 + 3z_0 + \frac{3}{z_0} + \left(\frac{1}{z_0}\right)^3$ and also that $\left(z_0 + \frac{1}{z_0}\right)^2 = z_0^2 + 2 + \left(\frac{1}{z_0}\right)^2$. It follows that

$$z_0^3 + z_0^2 + z_0 + 1 + \frac{1}{z_0} + \frac{1}{z_0^2} + \frac{1}{z_0^3} = \left(z_0 + \frac{1}{z_0}\right)^3 + \left(z_0 + \frac{1}{z_0}\right)^2 - 2\left(z_0 + \frac{1}{z_0}\right) - 1$$

Then, since $x_0 = z_0 + \frac{1}{z_0}$, x_0 satisfies the equation

$$x_0^3 + x_0^2 - 2x_0 - 1 = 0$$

As indicated, to show that a regular polygon with 7 sides is not constructible, it suffices to show that x_0 is not a constructible number. Since x_0 satisfies this cubic equation with rational coefficients, the result will follow if it is shown that this cubic equation has no rational root (Theorem 12.3.22). Suppose $\frac{m}{n}$ is a rational root written in lowest terms. Then, by the Rational Roots Theorem (8.2.14), m divides -1 and n divides 1. Therefore, m is 1 or -1, and n is 1 or -1. Thus, $\frac{m}{n}$ equals 1 or -1. But $1^3 + 1^2 - 2 - 1$ is not 0, nor is $(-1)^3 + (-1)^2 + 2 - 1$. Hence, there is no rational solution, and the theorem is proven. $\square$

In a sense, it is known exactly which regular polygons are constructible. The Gauss-Wantzel Theorem (which we will not prove) states that a regular polygon with n sides is constructible if and only if n is 2^k, where k is an integer greater than 1, or n is $2^k F_1 \cdots F_l$, where k is a nonnegative integer and the F_j are distinct Fermat primes. Recall (Problem 14 in Chapter 2) that a Fermat number is a number of the form $2^{2^n} + 1$ for a nonnegative integer n. A *Fermat prime* is a Fermat number that is prime. The first few Fermat numbers are 3 (when $n = 0$), 5 (when $n = 1$), 17 (when $n = 2$), and 257 (when $n = 3$). Fermat thought that all Fermat numbers might be prime, but Euler found that the sixth Fermat number (when $n = 5$) is not prime. It is remarkable fact that it is still unknown whether or not there are an infinite number of Fermat primes. (It is equally remarkable that it is not known whether there are infinitely many composite Fermat numbers.) It is therefore not known whether or not there are an infinite number of constructible regular polygons with an odd number of sides.

We can determine exactly which angles having a natural number of degrees are constructible.

Theorem 12.4.13. *If n is a natural number, then an angle of n degrees is constructible if and only if n is a multiple of 3.*

Proof. Recall that we proved that a regular polygon with 10 sides is constructible (Theorem 12.4.10) and, hence, that an angle of $36°$ is constructible (Theorem 12.4.5). Since an angle of $30°$ is constructible (Corollary 12.3.15), we can "subtract" a $30°$ angle from a $36°$ angle by placing the $30°$ angle with the vertex and one of its sides coincident with the vertex and one of the sides of the $36°$ angle (Theorem 12.1.6). Then, bisecting the constructed angle of $6°$ yields an angle of $3°$. Once an angle of $3°$ is constructed, an angle of $3k$ degrees can be constructed by simply constructing k angles of $3°$ appropriately adjacent to each other.

To establish the converse, suppose that an angle of n degrees is constructible. We must show that n is congruent to 0 (mod 3). If n were congruent to either 1 or 2 modulo 3, then we could construct an angle of $1°$ or $2°$ accordingly by "subtracting" an appropriate number of angles of $3°$ from the angle of n degrees. If the resulting angle is $2°$, bisecting it would yield an angle of $1°$. Thus, if an angle of n degrees was constructible for any n that was not a multiple of 3, then an angle of $1°$ could be constructed. But an angle of $1°$ is not constructible, for, if it was, placing 20 of them together would contradict the fact that an angle of $20°$ is not constructible (Theorem 12.3.23). □

We have shown that some angles, such as an angle of $60°$, cannot be trisected with a straightedge and compass. But what about the following?

Example 12.4.14 (Trisection of arbitrary acute angles). Let θ be any acute angle. Mark any two points on a straightedge and let the distance between them be r. Given the angle θ, construct the circle with radius r whose center is at the vertex of θ. Label the center of the circle O. Extend one of the sides of θ in both directions. Move the marked straightedge so that the point marked to the left is on the extended line, the point marked to the right stays on the circle, and the straightedge passes through the intersection of the circle and the side of θ that was not extended; label the points of intersection A, B, C, as shown in Figure 12.16.

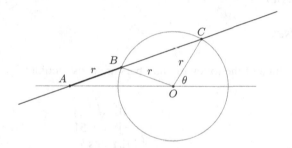

Fig. 12.16 On the way to trisecting an arbitrary angle θ

Draw the line BO. Then the line segments AB, BO, and OC all have length r. Now let the equal base angles of $\triangle ABO$ be x, the equal base angles of $\triangle OBC$ be y, and let $\angle BOC$ be z, as shown in Figure 12.17. Then the sum of $\angle ABO$ and $2x$ is $180°$, and the sum of $\angle ABO$ and y is also $180°$; hence $y = 2x$. It is clear that $x + z + \theta$ is $180°$. On the other hand, $z + 2y$ is $180°$. Since $y = 2x$, $4x + z$ is also $180°$. It follows that $4x + z = x + z + \theta$, or $3x = \theta$. Thus, the angle x is one third of θ, so θ has been trisected. □

What is going on here? You may think that the construction we have just done contradicts our earlier proof that an angle of $60°$ cannot be trisected. However, the construction in the example above violated the classical rules of constructions that we were adhering to before this example. Namely, we marked two points on the straightedge.

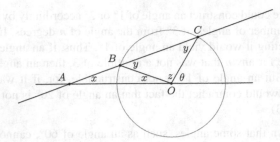

Fig. 12.17 Trisecting an arbitrary angle θ

What we have shown is that it *is* possible to trisect arbitrary angles with a compass and a straightedge on which two (or more) points are marked. Therefore, in particular, any angle can be trisected using a *ruler* and compass, but not merely using a straightedge and compass.

12.5 Problems

Basic Exercises

1. Determine which of the following numbers are constructible:

(a) $\dfrac{1}{\sqrt{3+\sqrt{2}}}$

(b) $\sqrt[3]{79}$

(c) 3.146891

(d) $\sqrt[16]{79}$

(e) $\sqrt{6+\frac{\sqrt[3]{4}}{2}}$

(f) $\sqrt{7+\sqrt{5}}$

(g) $\sqrt{3+4\sqrt{2}+\sqrt{5}}$

(h) $\sqrt[3]{\frac{9}{10}}$

(i) $\sqrt{\frac{3792}{1419}}$

(j) $\cos 51°$

(k) $\cos 5°$

(l) $\cos 10°$

(m) $11^{\frac{2}{3}}$

(n) $11^{\frac{3}{2}}$

(o) $2^{\frac{1}{6}}$

(p) $2^{\frac{3}{2}}$

(q) $\sqrt[3]{\frac{\sqrt{2}}{4}}$ [Hint: Consider its square.]

(r) $\sqrt{7}\cos 15°$

2. Determine which of the following angles are constructible:

(a) $6°$

(b) $5°$

(c) $10°$

(d) $30°$

(e) $35°$

(f) $15°$

(g) $75°$

(h) $80°$

(i) $92.5°$

(j) $37.5°$

(k) $7.5°$

(l) $120°$

(m) $160°$

3. Determine which of the following angles can be trisected:

    (a) $12°$
    (b) $30°$

Interesting Problems

4. Determine which of the following polynomials have at least one constructible root:

    (a) $x^4 - 3$ (f) $x^3 - 2x - 1$
    (b) $x^8 - 7$ (g) $x^3 + 4x + 1$
    (c) $x^4 + \sqrt{7}x^2 - \sqrt{3} - 1$ (h) $x^3 + 2x^2 - x - 1$
    (d) $x^3 + 6x^2 + 9x - 10$ (i) $x^3 - x^2 + x - 1$
    (e) $x^3 - 3x^2 - 2x + 6$ (j) $2x^3 - 4x^2 + 1$

5. Determine which of the following regular polygons can be constructed with straightedge and compass:

    (a) A regular polygon with 14 sides
    (b) A regular polygon with 20 sides
    (c) A regular polygon with 36 sides
    (d) A regular polygon with 240 sides

6. Explain how to construct a regular polygon with 24 sides using straightedge and compass.

7. True or False:

    (a) If the angle of θ degrees is constructible and the number x is constructible, then the angle of $x \cdot \theta$ degrees is constructible.
    (b) x^y is constructible if x and y are each constructible.
    (c) If $\frac{x}{z}$ is constructible, then x and z are each constructible.
    (d) There is an angle θ such that $\cos \theta$ is constructible but $\sin \theta$ is not constructible.

8. For an acute angle θ, show that $\tan \theta$ is a constructible number if and only if θ is a constructible angle.

9. Determine which of the following numbers are constructible:

    (a) $\sin 20°$
    (b) $\sin 75°$
    (c) $\tan 2.5°$

10. Determine which of the following numbers are constructible (the angles below are in radians):

    (a) $\sin \frac{\pi}{16}$

(b) $\cos \pi$

(c) $\tan \frac{\pi}{4}$

11. (a) Prove that the cube cannot be tripled, in the sense that, starting with an edge of a cube of volume 1, an edge of a cube of volume 3 cannot be constructed with straightedge and compass.

   (b) More generally, prove that the side of a cube with volume a natural number n is constructible if and only if $n^{\frac{1}{3}}$ is a natural number.

12. Using mathematical induction, prove that, for every integer $n \geq 1$, a regular polygon with $3 \cdot 2^n$ sides can be constructed with straightedge and compass.

Challenging Problems

13. Prove that the center of any given regular polygon can be constructed using only a straightedge and compass.

   [Hint: The center can be determined as the point of intersection of the perpendicular bisectors of two adjacent sides of the polygon. To prove that this point is indeed the center, prove that all the right triangles with one side a perpendicular bisector of a side of the polygon, another side a half of a side of the polygon, and the third side the line segment joining the "center" to a vertex of the polygon are congruent to each other.]

14. Prove that an acute angle cannot be trisected with straightedge and compass if its cosine is:

   (a) $\frac{3}{7}$ (d) $\frac{3}{5}$

   (b) $\frac{2}{5}$ (e) $\frac{1}{4}$

   (c) $\frac{1}{5}$

15. Can a polynomial of degree 4 with rational coefficients have a constructible root without having a rational root?

16. Prove that the following equation has no constructible solutions:

$$x^3 - 6x + 2\sqrt{2} = 0$$

   [Hint: You can use Theorem 12.3.22 if you make an appropriate substitution.]

17. Let t be a transcendental number. Prove that $\{(a + bt) : a, b \in \mathbb{Q}\}$ is not a subfield of $\mathbb{R}$.

18. Say that a complex number $a + bi$ is constructible if the point (a, b) is constructible (equivalently, if a and b are both constructible real numbers). Show that the cube roots of $\frac{1}{2} + \frac{\sqrt{3}}{2}i$ are not constructible.

19. Let $\mathcal{F}$ be the smallest subfield of $\mathbb{R}$ that contains π.

(a) Show that $\mathcal{F}$ consists of all numbers that can be written in the form $\frac{p(\pi)}{q(\pi)}$, where p and q are polynomials with rational coefficients and q is not the zero polynomial.

(b) Show that $\mathcal{F}$ is countable.

20. Is $\left\{ a\sqrt{2} : a \in \mathbb{Q} \right\}$ a subfield of $\mathbb{R}$?

21. Is the set of all towers countable? (Recall that a *tower* is a finite sequence of subfields of $\mathbb{R}$, the first of which is $\mathbb{Q}$, such that the other subfields are obtained from their predecessors by adjoining square roots; see Definition 12.2.16.)

22. Prove the following:

(a) If x_0 is a root of a polynomial with coefficients in $\mathcal{F}\left(\sqrt{r}\right)$, then x_0 is a root of a polynomial with coefficients in $\mathcal{F}$.

(b) Every constructible number is algebraic.

(c) The set of constructible numbers is countable.

(d) There is a circle with center at the origin that is not constructible.

23. Let t be a transcendental number. Prove that t cannot be a root of any equation of the form $x^2 + ax + b = 0$ if a and b are constructible numbers.

24. Is there a line in the plane such that every point on it is constructible?

25. Find the cardinality of each of the following sets:

(a) The set of roots of polynomials with constructible coefficients

(b) The set of constructible angles

(c) The set of all points (x, y) in the plane such that x is constructible and y is irrational

(d) The set of all sets of constructible numbers

26. (Very challenging) Use a straightedge and compass to directly (without first constructing its central angle or the length of the side of any polygon) construct a regular pentagon.

27. Suppose that regular polygons with m sides and n sides can be constructed and m and n are relatively prime. Prove that a regular polygon of mn sides can be constructed.

[Hint: Use central angles and use the fact that a linear combination of m and n is 1.]

28. Prove the following: For natural numbers m and n, if a given angle can be divided into n equal parts using only a straightedge and compass, and if m is a divisor of n, then the angle can be divided into m equal parts using only a straightedge and compass.

29. (Very challenging) Prove that you cannot trisect an angle by trisecting the side
    opposite the angle in a triangle containing it. That is, prove that, if ABC is any
    triangle, there do not exist two lines through A such that those lines trisect both
    the side BC of the triangle and the angle BAC of the triangle.

    [Hint: Suppose that there do exist two such lines. The lines then divide the
    triangle into three sub-triangles. One approach uses the easily established fact
    that all three sub-triangles have the same area.]

Chapter 13
An Introduction to Infinite Series

What is $.3333\ldots$ (where the sequence of 3's continues forever)? Presumably, this expression means

$$\frac{3}{10} + \frac{3}{10^2} + \frac{3}{10^3} + \frac{3}{10^4} + \cdots$$

where the "$\cdots$" indicates that the "sum" continues indefinitely. But then, what does *that* mean? What does it mean to add up an infinite number of terms? Is the sum $\frac{1}{3}$? Or is it merely close to $\frac{1}{3}$? Does it have a precise meaning?

Similarly, what is

$$\frac{1}{2} + \frac{1}{4} + \frac{1}{8} + \frac{1}{16} + \cdots$$

(where the indicated sum continues forever)? Expressions such as these are called "infinite series." In some cases, as we shall see, there is a natural way of defining the sum of an infinite series.

In this chapter, we present the basic facts about infinite series, emphasizing an understanding of the fundamental concepts. In order to make the central idea more accessible, our approach is a little unorthodox; the more standard approach is explained in Problem 27 at the end of the chapter.

13.1 Convergence

The rough idea is to define the sum of an infinite series to be S if adding any large enough (but finite) number of terms produces a sum that is very close to S. We will have to make precise what is meant by "very close." Before we present the formal definitions, we discuss the following example:

© Springer Nature Switzerland AG 2018
D. Rosenthal et al., *A Readable Introduction to Real Mathematics*,
Undergraduate Texts in Mathematics, https://doi.org/10.1007/978-3-030-00632-7_13

$$\frac{1}{2} + \frac{1}{4} + \frac{1}{8} + \frac{1}{16} + \cdots$$

We know what is meant by $\frac{1}{2} + \frac{1}{4}$; it is $\frac{3}{4}$. We know what is meant by $\frac{1}{2} + \frac{1}{4} + \frac{1}{8}$; it is $\frac{7}{8}$. Similarly, we know what is meant by $\frac{1}{2} + \frac{1}{4} + \frac{1}{8} + \frac{1}{16}$, or by the sum of any finite number of terms of this series. The sum of all the infinitely many terms is defined as a kind of *limit*, as we now explain.

For each natural number n, let $S_n = \frac{1}{2} + \frac{1}{4} + \frac{1}{8} + \cdots + \frac{1}{2^n}$ denote the sum of the first n terms of the series. We seek a simpler formula for S_n. Multiplying both sides of the equation defining S_n by $\frac{1}{2}$ yields $\frac{1}{2}S_n = \frac{1}{4} + \frac{1}{8} + \frac{1}{16} + \cdots + \frac{1}{2^{n+1}}$. Therefore,

$$S_n - \frac{1}{2}S_n = \left(\frac{1}{2} + \frac{1}{4} + \frac{1}{8} + \cdots + \frac{1}{2^n}\right) - \left(\frac{1}{4} + \frac{1}{8} + \frac{1}{16} + \cdots + \frac{1}{2^{n+1}}\right)$$

$$= \frac{1}{2} - \frac{1}{2^{n+1}}$$

In other words, $\frac{1}{2}S_n = \frac{1}{2} - \frac{1}{2^{n+1}}$. Thus, $S_n = 2\left(\frac{1}{2} - \frac{1}{2^{n+1}}\right) = 1 - \frac{1}{2^n}$.

This formula shows that S_n is close to 1 if n is large. In fact, no matter how close we want S_n to be to 1, we can guarantee that S_n will be at least that close by choosing n sufficiently large. To see this, note that the difference between S_n and 1 is $\frac{1}{2^n}$. Thus, for example, we can guarantee that S_n is within $\frac{1}{10}$ of 1 by taking n to be greater than or equal to 4, since $\frac{1}{16}$ is less than $\frac{1}{10}$. We can guarantee that S_n is within $\frac{1}{1000}$ of 1 by choosing n to be greater than or equal to 10, since $\frac{1}{2^{10}}$ (that is, $\frac{1}{1024}$) is less than $\frac{1}{1000}$. Since S_n is very close to 1 when n is large, the definition of the sum of an infinite series that we give below (Definition 13.1.4) implies that the sum of this infinite series is 1.

To define precisely what we mean by the sum of an infinite series, it is useful to have a notation for the distance between two real numbers. The following definition will play a central role.

Definition 13.1.1. The *absolute value* of a real number a, denoted $|a|$, is defined to be the distance from a to 0 on the number line. More precisely, if a is positive or zero, then $|a|$ is just a, and if a is negative, then $|a|$ is equal to $-a$.

For example, $|1| = 1$, $|-5| = -(-5) = 5$, $\left|-\frac{3}{\sqrt{2}}\right| = -\left(-\frac{3}{\sqrt{2}}\right) = \frac{3}{\sqrt{2}}$, and $|0| = 0$. Therefore, the absolute value has the effect of "making a number nonnegative."

Fig. 13.1 Solutions to $|a| = 5$

Consider the question: "Which real numbers a satisfy $|a| = 5$?" That is, "Which real numbers a have a distance of 5 from 0?" As can been seen by looking at the real line (as in Figure 13.1), there are only two directions one can move away from zero: the positive direction and the negative direction. Thus, there are only two real numbers that have a distance of 5 from 0; namely, 5 and -5. More generally, for every positive real number d, there are precisely two real numbers with absolute value equal to d; namely, d and $-d$.

What does $|a| \leq 5$ mean? This means that a has a distance from 0 that is less than or equal to 5. Looking at the real number line, we see that this means that a cannot be larger than 5 and also cannot be less than -5. In other words, $|a| \leq 5$ is equivalent to $-5 \leq a \leq 5$. More generally, if $d > 0$, then $|a| \leq d$ is equivalent to $-d \leq a \leq d$.

The distance between two numbers a and b can be expressed using absolute values as $|a - b|$. We show this as follows. Since this is clearly true when a or b is zero, we need only consider the cases where a and b are both nonzero. When a and b are both positive numbers it is not hard to see that the distance between them is $|a - b|$; simply picture a and b on a number line, and observe that, regardless of which is larger, the larger minus the smaller is $|a - b|$ (which is the same as $|b - a|$). For example, the distance between 2 and 5 is 3, and whether we write $|5 - 2|$ or $|2 - 5|$, we get 3. The case where a and b are both negative numbers can be obtained from the previous case by noting that the distance between a and b is the same as the distance between $-a$ and $-b$, which is $|(-a) - (-b)| = |-a + b| = |-(a - b)| = |a - b|$.

The case that takes a little more thought is the case where one of a and b is positive and the other is negative. Suppose that a is negative and b is positive (the other way around is handled analogously). Looking at the number line (Figure 13.2), since 0 is between a and b, we see that the distance between a and b is equal to the distance from a to 0 plus the distance from b to 0. That is, the distance between a and b is equal to $|a| + |b|$. Since $a < 0$ and $b > 0$, $|a| + |b|$ is equal to $(-a) + b = b - a = |b - a| = |a - b|$. Thus, the distance between a and b is $|a - b|$ in all cases.

$$a \qquad\qquad 0 \qquad\qquad\qquad\qquad b$$

Fig. 13.2 Distance between two real numbers

Using absolute values to denote distances between numbers is useful in formulating the precise definition of what is meant by the sum of an infinite series.

We return to the example $\frac{1}{2} + \frac{1}{4} + \frac{1}{8} + \cdots$. We want to capture the idea that, no matter how close we require the sum $S_n = \frac{1}{2} + \frac{1}{4} + \frac{1}{8} + \cdots + \frac{1}{2^n}$ to be to 1, we can get it that close by including a sufficient number of terms; i.e., by choosing n to be sufficiently large.

It is common in mathematics to use the Greek letter epsilon, written in the form ε, to denote a small positive real number, as in the following. For every positive number ε, there is a natural number N such that S_n is within ε of 1 whenever n is greater than or equal to N. This can be reformulated in terms of absolute values as the statement: For every $\varepsilon > 0$ there is a natural number N such that $|S_n - 1| < \varepsilon$ whenever $n \geq N$. This precisely captures what is meant by the requirement that S_n gets arbitrarily close to 1 when n is sufficiently large.

To see that this precise statement holds for $S_n = \frac{1}{2} + \frac{1}{4} + \frac{1}{8} + \cdots + \frac{1}{2^n} = 1 - \frac{1}{2^n}$, first note that $|S_n - 1| = \left| -\frac{1}{2^n} \right| = \frac{1}{2^n}$. Therefore, we must show that, for every $\varepsilon > 0$, there exists a natural number N such that $\frac{1}{2^n} < \varepsilon$ whenever $n \geq N$.

Let ε be any positive number. We find such an N by first noting that, since powers of 2 can be arbitrarily large, there is an N such that $2^N > \frac{1}{\varepsilon}$. Fix such an N. Multiplying both sides of $\frac{1}{\varepsilon} < 2^N$ by $\frac{\varepsilon}{2^N}$ gives $\frac{1}{2^N} < \varepsilon$. If $n \geq N$, then $\frac{1}{2^n} \leq \frac{1}{2^N} < \varepsilon$. Therefore, for each $\varepsilon > 0$, if N is large enough that 2^N is greater than $\frac{1}{\varepsilon}$, then S_n is within ε of 1 whenever $n \geq N$, as desired. This proves that, according to the general definition we give below (13.1.4), the sum of the infinite series $\frac{1}{2} + \frac{1}{4} + \frac{1}{8} + \frac{1}{16} + \cdots$ is 1.

Definition 13.1.2. An *infinite series* is an expression of the form

$$a_1 + a_2 + a_3 + \cdots$$

where the a_i are real numbers and the indicated sum continues forever. The a_i are called the *terms* of the series.

Some examples of infinite series are:

$$1 + \frac{1}{2} + \frac{1}{3} + \frac{1}{4} + \cdots$$

$$1 + \frac{1}{2} - \frac{1}{3} + \frac{1}{4} - \cdots$$

$$1 + \frac{1}{\sqrt{2}} + \frac{1}{\sqrt{3}} + \frac{1}{\sqrt{4}} + \cdots$$

$$1 - 1 + 1 - 1 + \cdots$$

$$79^2 + 80^2 + 81^2 + 82^2 + \cdots$$

$$\frac{2}{3} + \frac{2^2}{3^2} + \frac{2^3}{3^3} + \cdots$$

$$\frac{4}{12^3} + \frac{4}{13^3} + \frac{4}{14^3} + \frac{4}{15^3} + \cdots$$

Any particular infinite series may or may not have a *sum* according to the definition given below. The general definition is the same as the precise formulation in the special case considered above where $S_n = 1 - \frac{1}{2^n}$. The definition captures the rough idea that adding a large but finite number of terms of the series gives a partial sum that is close to the "sum" of the entire series.

Definition 13.1.3. For an infinite series

$$a_1 + a_2 + a_3 + a_4 + \cdots$$

the nth *partial sum*, often denoted by S_n, is the sum of the first n terms. That is,

$$S_n = a_1 + a_2 + a_3 + \cdots + a_n$$

Definition 13.1.4. The infinite series with partial sums S_n *converges to* S (or *has sum S*) if, for every $\varepsilon > 0$, there is a natural number N such that $|S_n - S| < \varepsilon$ whenever $n \geq N$.

As we showed above, the series $\frac{1}{2} + \frac{1}{4} + \frac{1}{8} + \cdots$ converges to 1 according to this definition.

Note that if there is any N that satisfies the definition for a particular ε, then there are infinitely many such N, for if N_0 satisfies the definition, so does any N larger than N_0.

Note also that a series does not converge to a given S if there is *any* ε greater than 0 for which there is no N satisfying the definition. (Of course, if ε_0 is such an ε, then so is any smaller positive number.)

For some infinite series, such as the example above where $S = 1$, it is not hard to determine the sum. There are other infinite series that can be shown to converge but for which finding an expression for the sum (other than the infinite series itself) is very difficult or even impossible. Moreover, there are infinite series that do not converge to any sum at all.

Definition 13.1.5. An infinite series *converges* if there is some S for which the series converges to S according to Definition 13.1.4. If a series does not converge to any S, we say that the series *diverges*.

Example 13.1.6. The series $1 + 1 + 1 + \cdots$ diverges.

Proof. The partial sums of this series are: $S_1 = 1$, $S_2 = 2$, $S_3 = 3$, and, for each natural number n, $S_n = n$. We want to show that there is no S that satisfies the definition of sum (13.1.4) for this series. Let S be any real number. If n is sufficiently large, then S_n will be much larger than S. To show that S is not the sum of the series, it suffices to find any $\varepsilon > 0$ for which there is no N satisfying Definition 13.1.4. For this particular series, in fact, every $\varepsilon > 0$ has that property. For instance, take $\varepsilon = 3$. No matter what N is chosen, there are an infinite number of n's greater than N for which S_n is greater than $S + 3$. For those S_n, it is not the case that $|S_n - S| < 3$. Therefore, the series diverges.   □

Example 13.1.7. The series $1 - 1 + 1 - 1 + 1 - 1 + \cdots$ diverges.

Proof. Consider the partial sums of this series: $S_1 = 1$, $S_2 = 0$, $S_3 = 1$, and so on. That is, the odd partial sums are all 1 and the even partial sums are all 0. To show that there is no S to which the series converges, it suffices to find some $\varepsilon > 0$ for which there is no N satisfying Definition 13.1.4. In this example, $\varepsilon = \frac{1}{3}$ (for instance) has that property. For, no matter what N is chosen, there will exist an $n_1 > N$ (any odd $n_1 > N$) and an $n_2 > N$ (any even $n_2 > N$) such that $S_{n_1} = 1$ and $S_{n_2} = 0$. There is no real number S that is within $\frac{1}{3}$ of both 1 and 0. Thus, the series diverges.   □

The proofs of many results concerning infinite series require an important inequality concerning absolute values. It is called the "triangle inequality" because its generalization to vectors in the plane is equivalent to the fact that the sum of the lengths of two sides of a triangle is greater than or equal to the length of the third side (see Theorem 14.4.9 in Chapter 14).

The Triangle Inequality 13.1.8. *Let x and y be real numbers. Then*

$$|x + y| \le |x| + |y|$$

Proof. Recall from our discussion about absolute values that if $d > 0$, then $|a| \le d$ is equivalent to $-d \le a \le d$. Therefore, we can prove the Triangle Inequality by showing that $-(|x| + |y|) \le x + y \le |x| + |y|$. For this, observe that for every real number a, $-|a| \le a \le |a|$. Thus, in particular, $-|x| \le x \le |x|$ and $-|y| \le y \le |y|$. Adding these two inequalities gives us the desired inequality, $-(|x| + |y|) \le x + y \le |x| + |y|$.   □

A fundamental question is: can a series have two different sums? That is, can a series converge to S and also converge to T, with S different from T? Our first application of the triangle inequality is in providing an answer to this question.

Theorem 13.1.9. *An infinite series converges to at most one real number. That is, if a series converges to S and also converges to T, then $S = T$.*

Proof. Let S_n be the nth partial sum of a given series, and suppose that the series converges to S and also to T. We will show that, for every $\varepsilon > 0$, $|S - T| < \varepsilon$. Since 0 is the only nonnegative real number that is less than every positive real number, it will then follow that $|S - T| = 0$; that is, $S = T$.

Let $\varepsilon > 0$ be given. For every n, the Triangle Inequality (13.1.8) implies

$$|S - T| = |S - S_n + S_n - T| = |(S - S_n) + (S_n - T)| \le |S - S_n| + |S_n - T|$$

Since the series converges to S, for every $\varepsilon_1 > 0$ there is an N_1 such that $|S_n - S| < \varepsilon_1$ for every $n \ge N_1$. Similarly, since the series converges to T, for every $\varepsilon_2 > 0$ there is an N_2 such that $|S_n - T| < \varepsilon_2$ for every $n \ge N_2$. Choose $\varepsilon_1 = \frac{\varepsilon}{2}$ and $\varepsilon_2 = \frac{\varepsilon}{2}$. If N is the larger of N_1 and N_2, then both inequalities are satisfied for all $n \ge N$. Thus, for $n \ge N$,

$$|S - S_n| + |S_n - T| < \varepsilon_1 + \varepsilon_2 = \frac{\varepsilon}{2} + \frac{\varepsilon}{2} = \varepsilon$$

Therefore, $|S - T| < \varepsilon$, as desired.   □

Despite the previous theorem, rearranging the order of the terms of an infinite series sometimes produces a series with a different sum (see Problem 19). That is, the order in which the terms of an infinite series are added is important under certain conditions (see Problems 21 and 22). This surprising possibility is, of course, different from the situation when adding a finite number of numbers. It is therefore important to note that the definition of a particular infinite series depends not only on the terms of the series but also on the order in which the terms are arranged. However, if the terms of the series are all nonnegative, then rearranging the order of the terms does not affect the sum (see Problem 22).

13.2   Geometric Series

The example $\frac{1}{2} + \frac{1}{4} + \frac{1}{8} + \cdots$ that we considered above is a particular instance of a special kind of series that is easily dealt with.

Definition 13.2.1. A series $a_1 + a_2 + a_3 \cdots$ is called a *geometric series* if there is a number r, called the *ratio* of the series, such that each term of the series is obtained from the preceding term by multiplying by r. That is, there is a number r such that, if the first term of the series is a, then the second term is ar, the third term is ar^2, and so on. Such a series has the form

$$a + ar^2 + ar^3 + \cdots + ar^n + \cdots$$

for some real numbers a and r.

The series that we previously discussed,

$$\frac{1}{2} + \frac{1}{4} + \frac{1}{8} + \cdots$$

is a geometric series with first term $a = \frac{1}{2}$ and ratio $r = \frac{1}{2}$.

We begin the analysis of geometric series by finding a formula for the nth partial sum of a general geometric series.

Theorem 13.2.2. *If a is a real number and r is a real number other than 1, and if*

$$S_n = a + ar + ar^2 + ar^3 + \cdots + ar^{n-1}$$

then

$$S_n = \frac{a - ar^n}{1 - r}$$

Proof. The proof is similar to that of the special case that we discussed above (where $a = \frac{1}{2}$ and $r = \frac{1}{2}$). Since

$$rS_n = ar + ar^2 + \cdots + ar^n$$

it follows that $S_n - rS_n = a - ar^n$ (since all of the other terms cancel each other out when doing the subtraction). Thus, $(1 - r)S_n = a - ar^n$, or $S_n = \frac{a-ar^n}{1-r}$. (Note that this formula for S_n does not make sense when $r = 1$. Of course, S_n is simply na in this case.) □

We will use the above formula to determine which geometric series converge and to find the sum of a geometric series when it exists. We need the fact that if a number has absolute value less than 1, then its powers get arbitrarily small. That is, we need the following lemma. (Like ε, the Greek letter δ, read "delta," is often used to denote a small positive number.)

Lemma 13.2.3. *If $|r| < 1$, then for every positive number δ there is a natural number N such that $|r|^n$ is less than δ for all $n \geq N$.*

Proof. If $r = 0$, then $r^n = 0$ for all n, and therefore any N will do. If r is not 0, then $|r| < 1$ implies that $\frac{1}{|r|}$ is greater than 1. Define the positive number t by $t = \frac{1}{|r|} - 1$, so $\frac{1}{|r|} = 1 + t$. We need the fact that, for every natural number n, $(1 + t)^n$ is greater than or equal to $1 + nt$ (which can be proven easily by mathematical induction, as stated in Problem 5 of this chapter). Now let $\delta > 0$ be given. Choose any N that is large enough so that Nt is greater than $\frac{1}{\delta}$. Suppose that $n \geq N$. Then $nt \geq Nt > \frac{1}{\delta}$. Thus, $(1 + t)^n \geq (1 + nt) > 1 + \frac{1}{\delta} > \frac{1}{\delta}$, so $(1 + t)^n > \frac{1}{\delta}$, or $\left(\frac{1}{|r|}\right)^n = \frac{1}{|r|^n} > \frac{1}{\delta}$. This implies that $|r|^n < \delta$, as desired. □

Theorem 13.2.4. *If a is a real number and r is a real number with $|r| < 1$, then the geometric series $a + ar + ar^2 + \cdots + ar^{n-1} + \cdots$ converges to $\frac{a}{1-r}$.*

Proof. The nth partial sum of the series is $S_n = \frac{a-ar^n}{1-r}$ (Theorem 13.2.2). According to the precise definition of convergence (13.1.4), the theorem will be proven if we establish that for every $\varepsilon > 0$ there exists an N such that $\left|S_n - \frac{a}{1-r}\right| < \varepsilon$ whenever $n \geq N$. Note that the difference between $\frac{a}{1-r}$ and S_n is

$$\frac{a}{1-r} - \frac{a-ar^n}{1-r} = \frac{ar^n}{1-r}$$

Let $\varepsilon > 0$ be given. If $a = 0$, then $\frac{ar^n}{1-r} = 0$. Assume now that $a \neq 0$. By Lemma 13.2.3, there is an N such that for all $n \geq N$, $|r|^n$ is less than the positive number $\frac{1-r}{|a|} \cdot \varepsilon$. Then, for all $n \geq N$,

$$\left|S_n - \frac{a}{1-r}\right| = \left|\frac{ar^n}{1-r}\right| = \frac{|a|}{1-r} \cdot |r^n| < \frac{|a|}{1-r} \cdot \frac{1-r}{|a|} \cdot \varepsilon = \varepsilon$$

Therefore, $\frac{a}{1-r}$ is the sum of the series. □

The example with which we began is a special case of the above.

Example 13.2.5. $\frac{1}{2} + \frac{1}{4} + \frac{1}{8} + \cdots = 1$

Proof. This is the special case of Theorem 13.2.4 where $a = \frac{1}{2}$ and $r = \frac{1}{2}$. $\square$

There are, of course, many other geometric series.

Example 13.2.6. $\sqrt{17} + \frac{\sqrt{17}}{3} + \frac{\sqrt{17}}{3^2} + \frac{\sqrt{17}}{3^3} + \cdots = \frac{3\sqrt{17}}{2}$

Proof. This follows from Theorem 13.2.4 using $a = \sqrt{17}$ and $r = \frac{1}{3}$. The sum is $\frac{a}{1-r} = \frac{\sqrt{17}}{\left(\frac{2}{3}\right)} = \frac{3\sqrt{17}}{2}$. $\square$

Geometric series can have negative ratios.

Example 13.2.7. $1 - \frac{1}{2} + \frac{1}{4} - \frac{1}{8} + \frac{1}{16} - \cdots = \frac{2}{3}$

Proof. This is the case of Theorem 13.2.4 where $a = 1$ and $r = -\frac{1}{2}$. The sum is $S = \frac{1}{\left(1-(-\frac{1}{2})\right)} = \frac{1}{\left(\frac{3}{2}\right)} = \frac{2}{3}$. $\square$

The proper interpretation of $.333333\ldots$ is as an infinite series; it is another way of writing the series $\frac{3}{10} + \frac{3}{100} + \frac{3}{1000} + \cdots$. Hence, we can determine its value.

Example 13.2.8. The infinite decimal $.33333\ldots$ is $\frac{1}{3}$.

Proof. This follows from Theorem 13.2.4 with $a = \frac{3}{10}$ and $r = \frac{1}{10}$; the sum is $\frac{\left(\frac{3}{10}\right)}{\left(1-\frac{1}{10}\right)} = \frac{\left(\frac{3}{10}\right)}{\left(\frac{9}{10}\right)} = \frac{1}{3}$. $\square$

We shall see in Section 13.6 that every infinite decimal $.b_1 b_2 b_3 \ldots$, where each b_i is an integer between 0 and 9, is a convergent series (Theorem 13.6.2), although most infinite decimals are not geometric series.

13.3 Convergence of Related Series

In some cases the convergence of a series can be established by using the fact that a related series is known to converge.

Theorem 13.3.1. *If $a_1 + a_2 + a_3 + \cdots$ is an infinite series that converges to S and c is any real number, then the infinite series $ca_1 + ca_2 + ca_3 + \cdots$ (obtained by multiplying each of the terms of the original series by c) converges to cS.*

Proof. We want to show that the sum of the series $ca_1 + ca_2 + ca_3 + \cdots$ is cS. By Definition 13.1.4, we need to show that, for every $\varepsilon > 0$, there exists a natural number N such that the absolute value of the difference between cS and the nth partial sum of the series is less than ε for every $n \geq N$. In the case where $c = 0$, the series obviously converges to $0 \cdot S = 0$. We therefore assume that $c \neq 0$ in what follows.

Let S_n be the nth partial sum of the series $a_1 + a_2 + a_3 + \cdots$. Then the nth partial sum of the series $ca_1 + ca_2 + ca_3 + \cdots$ is cS_n. Therefore, we need to consider $|cS_n - cS|$. But this is equal to $|c(S_n - S)| = |c| \cdot |S_n - S|$. This enables us to prove the theorem, as follows. Let $\varepsilon > 0$ be given. Since $a_1 + a_2 + a_3 + \cdots$ converges to S, there is a natural number N such that S_n is within the positive number $\frac{\varepsilon}{|c|}$ of S for every $n \geq N$. That is, $|S_n - S| < \frac{\varepsilon}{|c|}$ for every $n \geq N$. Therefore, $|cS_n - cS| = |c| \cdot |S_n - S| < |c| \cdot \frac{\varepsilon}{|c|} = \varepsilon$ for every $n \geq N$, so Definition 13.1.4 is satisfied by cS.   □

Example 13.3.2. The infinite decimal $.66666\ldots$ is equal to $\frac{2}{3}$ and the infinite decimal $.99999\ldots$ is equal to 1.

Proof. We could establish both of these facts using Theorem 13.2.4, as we did in Example 13.2.8. But, since we already know that the infinite decimal $.33333\ldots$ is $\frac{1}{3}$, we can use the theorem we just proved (13.3.1) to get the result even more simply. The infinite decimal $.66666\ldots$ is twice the infinite decimal $.33333\ldots$. More precisely, each of the terms of the infinite series representing $.66666\ldots$ is equal to twice the corresponding term of the infinite series that represents $.33333\ldots$. Thus, by Theorem 13.3.1, $.66666\ldots$ is equal to $2 \cdot \frac{1}{3} = \frac{2}{3}$. Similarly, since $.99999\ldots$ is 3 times $.33333\ldots$, Theorem 13.3.1 implies that $.99999\ldots$ is $3 \cdot \frac{1}{3} = 1$.   □

Theorem 13.3.3. *If $a_1 + a_2 + a_3 + \cdots$ is an infinite series that converges to S and $b_1 + b_2 + b_3 + \cdots$ is an infinite series that converges to T, then the infinite series $(a_1 + b_1) + (a_2 + b_2) + (a_3 + b_3) + \cdots$ converges to $S + T$.*

Proof. We must show that the sum of the series $(a_1 + b_1) + (a_2 + b_2) + (a_3 + b_3) + \cdots$ is $S + T$. By Definition 13.1.4, we need to show that, for every $\varepsilon > 0$, there exists a natural number N such that the absolute value of the difference between $S + T$ and the nth partial sum of the series is less than ε for every $n \geq N$.

Let S_n be the nth partial sum of the series $a_1 + a_2 + a_3 + \cdots$, and let T_n be the nth partial sum of the series $b_1 + b_2 + b_3 + \cdots$. Then $S_n + T_n$ is the nth partial sum of the series $(a_1 + b_1) + (a_2 + b_2) + (a_3 + b_3) + \cdots$. Therefore, we need to analyze $|(S_n + T_n) - (S + T)|$. First, $|(S_n + T_n) - (S + T)| = |(S_n - S) + (T_n - T)|$. By the Triangle Inequality (13.1.8), $|(S_n - S) + (T_n - T)| \leq |S_n - S| + |T_n - T|$. Thus, $|(S_n + T_n) - (S + T)| \leq |S_n - S| + |T_n - T|$.

This inequality enables us to prove the theorem, as follows. Let $\varepsilon > 0$ be given. Since $a_1 + a_2 + a_3 + \cdots$ converges to S, there exists an integer N_1 such that $|S_n - S| < \frac{\varepsilon}{2}$ for every $n \geq N_1$. Since $b_1 + b_2 + b_3 + \cdots$ converges to T, there is an integer N_2 such that $|T_n - T| < \frac{\varepsilon}{2}$ for every $n \geq N_2$. Thus, if we let N be the larger of N_1 and N_2, then both of these inequalities hold whenever $n \geq N$. Therefore, $|(S_n + T_n) - (S + T)| \leq |S_n - S| + |T_n - T| < \frac{\varepsilon}{2} + \frac{\varepsilon}{2} = \varepsilon$ for all $n \geq N$.   □

The above theorem can be summarized "convergent series can be added term-by-term."

13.4 Least Upper Bounds

In order to investigate infinite series other than geometric series, we need to understand an important property of the real numbers called "The Least Upper Bound Property."

Definition 13.4.1. If S is a set of real numbers, then the real number t is an *upper bound* for S if t is greater than or equal to every element of S. That is, t is an upper bound of S if $x \le t$ for every x in S.

For example, 2 is an upper bound of the set $\left\{1, \frac{1}{2}, \frac{1}{3}, \dots\right\}$; -1 is an upper bound of $\{-1, -2, -3, \dots\}$; 5 is an upper bound of $\left\{3 - \frac{1}{2}, 3 - \frac{1}{3}, 3 - \frac{1}{4}, \dots\right\}$; and 28 is an upper bound of $\{x : x < \sqrt{2}\}$.

It is important to observe that if t is an upper bound of a set S, then every real number greater than t is also an upper bound of S. Thus, if a set S has an upper bound, it has infinitely many upper bounds. Some sets, however, such as $\{1, 2, 3, 4, 5, \dots\}$ and $\left\{\left(\sqrt{2}\right)^n : n \in \mathbb{N}\right\}$, do not have any upper bounds at all.

Definition 13.4.2. If S is a set of real numbers, then the real number t is a *least upper bound* of S if t is an upper bound of S and every upper bound of S is greater than or equal to t. That is, a least upper bound is a smallest upper bound.

For example, a least upper bound of $\left\{1, \frac{1}{2}, \frac{1}{3}, \dots\right\}$ is 1. This can be seen as follows. Since every number in the set is less than or equal to 1, 1 is an upper bound. If t is any upper bound, then t is greater than or equal to 1 (since 1 is in the set), so 1 is a least upper bound.

A least upper bound of $\{-1, -2, -3, -4, -5, \dots\}$ is -1. To see this, first observe that every number in the set is less than or equal to -1, so -1 is an upper bound. Every upper bound must be greater than or equal to -1 since -1 is in the set.

A least upper bound of $\left\{3 - \frac{1}{2}, 3 - \frac{1}{3}, 3 - \frac{1}{4}, \dots\right\}$ is 3. To verify this, begin by noting that, since 3 is greater than any number in the set, 3 is an upper bound. To see that it is a least upper bound, note that if t is any upper bound, then t must be greater than or equal to $3 - \frac{1}{n}$ for every natural number n. If t was less than 3, then there would be some n such that $t < 3 - \frac{1}{n}$ (just choose n such that $\frac{1}{n} < 3 - t$). Thus, t would not be an upper bound. Therefore, all upper bounds are greater than or equal to 3, so 3 is a least upper bound. Note that, in this example, 3 is a least upper bound of the set but is not in the set.

A least upper bound of $\{x : x < \sqrt{2}\}$ is $\sqrt{2}$. To see this, note that every number in the set is less than $\sqrt{2}$, so $\sqrt{2}$ is an upper bound. If t is a number less than $\sqrt{2}$, then $t + \frac{1}{2}\left(\sqrt{2} - t\right) = \frac{1}{2}(t + \sqrt{2}) < \frac{1}{2}(\sqrt{2} + \sqrt{2}) = \sqrt{2}$. Therefore $t + \frac{1}{2}\left(\sqrt{2} - t\right)$ is in the set. But $t + \frac{1}{2}\left(\sqrt{2} - t\right)$ is obviously greater than t, since $\sqrt{2} - t$ is positive. Therefore such a t is not an upper bound for the set. Thus, $\sqrt{2}$ is a least upper bound.

Suppose that S is the empty set; that is, the set that does not contain any elements. Then every real number is an upper bound for S, for, no matter what real number is given, S does not contain any numbers greater than it. Since every real number is an upper bound for the empty set, the empty set does not have a *least* upper bound.

A crucial property of the real numbers that we will need in order to understand infinite series, and which is also important in many other contexts, is the existence of least upper bounds. We assume as an axiom that the real numbers have this property. (In fact, this property can be proven from the Dedekind cuts construction of the real numbers in terms of sets of rational numbers; see Problem 26 at the end of this chapter.)

The Least Upper Bound Property 13.4.3. *Every nonempty set of real numbers that has an upper bound has a least upper bound. In other words, the set of upper bounds has a smallest element. More precisely, if S is any set other than the empty set and S has an upper bound, then there is an upper bound t_0 for S such that every upper bound for S is greater than or equal to t_0.*

We next show that a set cannot have two different least upper bounds.

Theorem 13.4.4. *If a nonempty set of real numbers has an upper bound, then the set has a unique least upper bound.*

Proof. By the Least Upper Bound Property (13.4.3), every nonempty set of real numbers that has an upper bound has a least upper bound; we must show that there is at most one least upper bound for a given set. Suppose that both t_1 and t_2 are least upper bounds of the same nonempty set. Then, since t_1 is a least upper bound and t_2 is another upper bound, $t_1 \leq t_2$. Similarly, since t_2 is a least upper bound and t_1 is another upper bound, $t_2 \leq t_1$. Therefore, $t_1 = t_2$.   □

Theorem 13.4.5. *If an infinite series converges, then the set of partial sums of the series has an upper bound.*

Proof. Suppose that an infinite series converges to the sum S. Then, by Definition 13.1.4, for every $\varepsilon > 0$ there exists an N such that $|S_n - S| < \varepsilon$ whenever $n \geq N$. In particular, there is such an N_0 for $\varepsilon = 1$. That is, the absolute value of $S_n - S$ is less than 1 whenever n is greater than or equal to N_0. It follows, in particular, that for all $n \geq N_0$, $S_n - S < 1$. Therefore, $S_n < 1 + S$ for every $n \geq N_0$. Thus, we have found an upper bound, $1 + S$, for the set of all partial sums except for the first $N_0 - 1$ of them. Let t be the largest of the numbers in the finite set $\{S_1, S_2, \ldots, S_{N_0}\}$. Then the larger of $1 + S$ and t is an upper bound for the set of all partial sums.   □

The converse of Theorem 13.4.5 is not true in general, as the next example shows.

Example 13.4.6. The set of partial sums of the series $1 - 1 + 1 - 1 + \cdots$ has an upper bound, but the series does not converge.

Proof. In Example 13.1.7, we showed that this series diverges. Since the set of partial sums is simply $\{0, 1\}$, 1 is an upper bound for the set of partial sums.   □

13.5 The Comparison Test

In the case where all of the terms of an infinite series are nonnegative, the converse of Theorem 13.4.5 is true.

Theorem 13.5.1. *If $a_1 + a_2 + a_3 + \cdots$ is an infinite series with $a_i \geq 0$ for all i, then the series converges if and only if the set of all partial sums has an upper bound.*

Proof. Convergence implies that the set of partial sums has an upper bound, by Theorem 13.4.5. The proof of the converse requires the Least Upper Bound Property (13.4.3). To begin the proof, assume that the set of partial sums has an upper bound. Then, by the Least Upper Bound Property, the set of partial sums has a least upper bound. Call this least upper bound S; we prove that the series converges to S. For this, we must show (Definition 13.1.4) that, for every $\varepsilon > 0$, there is an N such that $|S_n - S| < \varepsilon$ whenever $n \geq N$.

Let $\varepsilon > 0$ be given. Since ε is greater than 0, $S - \varepsilon < S$. Since S is the *least* upper bound of the set of partial sums and $S - \varepsilon$ is less than S, $S - \varepsilon$ cannot be an upper bound of the set of partial sums. Thus, there is some N such that $S - \varepsilon < S_N$. We show that this N satisfies the definition of convergence (13.1.4) for the given ε. The hypothesis that each term of the series is nonnegative implies that if $n_1 \leq n_2$, then $S_{n_1} \leq S_{n_2}$, since S_{n_2} is obtained by adding nonnegative numbers to S_{n_1}. In particular, $S_N \leq S_n$ whenever $n \geq N$. Also, since S is an upper bound for the set of partial sums, $S_n \leq S$ for every n. Therefore, when $n \geq N$,

$$S - \varepsilon < S_N \leq S_n \leq S$$

In particular, $S - \varepsilon < S_n < S + \varepsilon$ for every such n. Subtracting S from each term in this inequality yields $-\varepsilon < S_n - S < \varepsilon$, which is equivalent to $|S_n - S| < \varepsilon$. Thus, S_n is within ε of S for every $n \geq N$, and the theorem follows. $\square$

Example 13.5.2. The series

$$\frac{1}{1 \cdot 2} + \frac{1}{2 \cdot 2^2} + \frac{1}{3 \cdot 2^3} + \frac{1}{4 \cdot 2^4} + \cdots + \frac{1}{n \cdot 2^n} + \cdots$$

converges.

Proof. First, $\frac{1}{n \cdot 2^n}$ is less than or equal to $\frac{1}{2^n}$ for every natural number n. Thus, every partial sum of this series is less than or equal to the corresponding partial sum of the geometric series $\frac{1}{2} + \frac{1}{2^2} + \frac{1}{2^3} + \cdots$. Since the sum of the geometric series with ratio $\frac{1}{2}$ and first term $\frac{1}{2}$ is 1 (Example 13.2.5), it follows that the set of partial sums of both series have 1 as an upper bound. Therefore, by Theorem 13.5.1, the series converges. $\square$

The previous example is a special case of a general situation: A series with nonnegative terms must converge if it is "term by term" less than or equal to a convergent series. That is, the following theorem holds.

The Comparison Test 13.5.3. *Suppose* $0 \leq a_n \leq b_n$ *for every natural number* n.

(i) *If the series* $b_1 + b_2 + b_3 + \cdots$ *converges, then the series* $a_1 + a_2 + a_3 + \cdots$ *converges.*

(ii) *If the series* $a_1 + a_2 + a_3 + \cdots$ *diverges, then the series* $b_1 + b_2 + b_3 + \cdots$ *diverges.*

Proof. (i) Let S be the sum of the series $b_1 + b_2 + b_3 + \cdots$. It is clear that every partial sum of the series $a_1 + a_2 + a_3 + \cdots$ is at most S. Thus, by Theorem 13.5.1, the series $a_1 + a_2 + a_3 + \cdots$ converges.

(ii) If $b_1 + b_2 + b_3 + \cdots$ did converge, then $a_1 + a_2 + a_3 + \cdots$ would converge, by part (i). Therefore, $b_1 + b_2 + b_3 + \cdots$ diverges.   □

The following is a very easy application of the Comparison Test.

Example 13.5.4. The series $\frac{1}{1^2 \cdot 7} + \frac{1}{2^2 \cdot 7^2} + \frac{1}{3^2 \cdot 7^3} + \frac{1}{4^2 \cdot 7^4} + \cdots + \frac{1}{n^2 \cdot 7^n} + \cdots$ converges.

Proof. This series is clearly term by term less than the convergent geometric series $\frac{1}{7} + \frac{1}{7^2} + \frac{1}{7^3} + \cdots$.   □

A slightly more complicated application is the following.

Example 13.5.5. The series

$$\frac{1}{3} + \frac{2}{3^2} + \frac{3}{3^3} + \frac{4}{3^4} + \cdots + \frac{n}{3^n} + \cdots$$

converges.

Proof. We will establish that $\frac{n}{3^n}$ is less than $\frac{1}{2^n}$ for every natural number n; the convergence of the given series then follows from part (i) of the Comparison Test (13.5.3) by comparing the given series to the geometric series $\frac{1}{2} + \frac{1}{4} + \frac{1}{8} + \cdots$.

The fact that $\frac{n}{3^n}$ is less than $\frac{1}{2^n}$ for every natural number n can be established by the Generalized Principle of Mathematical Induction (2.1.4). To see this, note that $\frac{n}{3^n} < \frac{1}{2^n}$ is equivalent to $\left(\frac{2}{3}\right)^n < \frac{1}{n}$. This is true for $n = 1$ and also for $n = 2$. Now suppose that $\left(\frac{2}{3}\right)^k < \frac{1}{k}$ for some $k \geq 2$. Note that $\frac{2}{3} \leq \frac{k}{k+1}$ when $k \geq 2$. It follows that

$$\left(\frac{2}{3}\right)^{k+1} = \left(\frac{2}{3}\right)^k \cdot \frac{2}{3} < \frac{1}{k} \cdot \frac{k}{k+1} = \frac{1}{k+1}$$

Therefore, $\frac{n}{3^n} < \frac{1}{2^n}$ for all $n \geq 2$. Thus, part (i) of the Comparison Test (13.5.3) gives the result.   □

13.6 Representing Real Numbers by Infinite Decimals

We show that every infinite decimal represents a real number, and, conversely, that every real number has a representation as an infinite decimal.

We define nonnegative infinite decimals as follows.

Definition 13.6.1. A *nonnegative infinite decimal* is an expression of the form $M.a_1a_2a_3\ldots$, where M is a nonnegative integer and each a_i is a "digit" (i.e., a number in the set $\{0, 1, \ldots, 9\}$). We interpret such an expression as representing the infinite series $M + \frac{a_1}{10} + \frac{a_2}{10^2} + \cdots + \frac{a_k}{10^k} + \cdots$.

Theorem 13.6.2. *Every nonnegative infinite decimal converges.*

Proof. We must show that the infinite series $M + \frac{a_1}{10} + \frac{a_2}{10^2} + \cdots + \frac{a_k}{10^k} + \cdots$ converges whenever M is a nonnegative integer and each a_k is a digit. Since each a_k is a digit, $\frac{a_k}{10^k}$ is less than or equal to $\frac{9}{10^k}$, for every k. It follows that the infinite decimal is term by term less than or equal to the "comparison series"

$$M + \frac{9}{10} + \frac{9}{10^2} + \frac{9}{10^3} + \cdots + \frac{9}{10^k} + \cdots$$

This comparison series converges to $M + \dfrac{\left(\frac{9}{10}\right)}{\left(1-\frac{1}{10}\right)} = M+1$, since its partial sums are all of the form M plus a partial sum of the convergent geometric series $\frac{9}{10} + \frac{9}{10^2} + \frac{9}{10^3} + \cdots$. Therefore, part (i) of the Comparison Test (13.5.3) gives the result. $\square$

Negative infinite decimals can be defined as negations of nonnegative ones. For example, $-7.11\ldots = -(7.11\ldots)$. In general, when M is a nonnegative integer and each a_k is a digit, we define $-M.a_1a_2a_3\ldots$ to be $(-1)\cdot\left(M + \frac{a_1}{10} + \frac{a_2}{10^2} + \frac{a_3}{10^3} + \cdots\right)$.

Theorem 13.6.3. *Every real number can be represented as an infinite decimal.*

Proof. First, the infinite decimal $.0000\ldots$ represents the real number 0. We next consider the case where r is a positive real number. We can obtain a decimal representation for r as follows. Let M be the greatest integer that is less than or equal to r. Such an M will be 0 or a natural number. Note that $r - M$ is less than 1, since otherwise $M + 1$ would be an integer less than or equal to r. Let a_1 be the largest digit such that $\frac{a_1}{10}$ is less than or equal to $r - M$. Then $r - \left(M + \frac{a_1}{10}\right)$ is less than $\frac{1}{10}$. Let a_2 be the largest digit such that $\frac{a_2}{10^2}$ is less than or equal to $r - \left(M + \frac{a_1}{10}\right)$. Then $r - \left(M + \frac{a_1}{10} + \frac{a_2}{10^2}\right)$ is less than $\frac{1}{10^2}$; then let a_3 be the largest digit such that $\frac{a_3}{10^3}$ is less than $r - \left(M + \frac{a_1}{10} + \frac{a_2}{10^2}\right)$.

Continue constructing digits a_k in this manner. Then, for each k, the absolute value of $r - \left(M + \frac{a_1}{10} + \frac{a_2}{10^2} + \cdots + \frac{a_k}{10^k}\right)$ is less than $\frac{1}{10^k}$. We claim that the infinite series $M + \frac{a_1}{10} + \frac{a_2}{10^2} + \frac{a_3}{10^3} + \cdots$ converges to r. To see this, let $\varepsilon > 0$ be given and choose N large enough that 10^{N-1} is greater than $\frac{1}{\varepsilon}$. If S_n is the nth partial sum of the series $M + \frac{a_1}{10} + \frac{a_2}{10^2} + \frac{a_3}{10^3} + \cdots$ and $n \geq N$, then

$$|r - S_n| = \left| r - \left(M + \frac{a_1}{10} + \frac{a_2}{10^2} + \cdots + \frac{a_{n-1}}{10^{n-1}} \right) \right| < \frac{1}{10^{n-1}} \leq \frac{1}{10^{N-1}} < \varepsilon$$

Thus, the series converges to r.

If r is a negative real number, an infinite decimal representation of r can be obtained by multiplying an infinite decimal representation of $-r$ by -1. □

Thus, real numbers can be represented by infinite decimals. However, the representation of a real number by an infinite decimal is not necessarily unique. Some real numbers have two distinct representations. For example, summing the corresponding infinite series shows that $.299999\ldots$ equals $.300000\ldots$. More generally, every real number represented by an infinite decimal that ends with an infinite string of 9's also has a representation that ends with a string of 0's. That is the only way that a real number can have two distinct infinite decimal representations. Before proving this fact, we make the following observation.

Lemma 13.6.4. *The infinite decimal* $.c_1c_2c_3\ldots$ *is at most 1. Moreover, if any* c_k *is less than 9, then* $.c_1c_2c_3\ldots$ *is less than 1. In particular,* $.c_1c_2c_3\ldots$ *is equal to 1 if and only if* $c_k = 9$ *for every* k.

Proof. The infinite series that $.c_1c_2c_3\ldots$ represents is term by term less than or equal to the geometric series $\frac{9}{10} + \frac{9}{10^2} + \frac{9}{10^3} + \cdots$. Since that geometric series sums to 1, every partial sum of $\frac{c_1}{10} + \frac{c_2}{10^2} + \frac{c_3}{10^3} + \cdots$ is at most 1. Therefore, the infinite decimal $.c_1c_2c_3\ldots$ is at most 1. If c_k is strictly less than 9, then every partial sum of $\frac{c_1}{10} + \frac{c_2}{10^2} + \frac{c_3}{10^3} + \cdots$ is less than $1 - \frac{1}{10^k}$, so $.c_1c_2c_3\ldots$ is at most $1 - \frac{1}{10^k}$. □

Theorem 13.6.5. *If two different infinite decimals represent the same real number, then one of them ends in a string of 9's and the other ends in a string of 0's.*

Proof. Clearly, it suffices to prove the theorem for representations of positive real numbers. Suppose, then, that a positive number has distinct representations

$$M.a_1a_2a_3\ldots = N.b_1b_2b_3\ldots$$

(Saying that the representations are distinct means that they differ in at least one digit.) Consider the case where M and N are different. Since $M \neq N$, one of them is larger. Suppose that M is greater than N; that is, $M - N \geq 1$. On the other hand, $M - N \leq 1$ since

$$M - N \leq (M - N) + .a_1a_2a_3\ldots = M.a_1a_2a_3\ldots - N = .b_1b_2b_3\ldots$$

which we know from Lemma 13.6.4 is at most 1. Thus, $M - N = 1$, and the above gives

$$1 \leq 1 + .a_1a_2a_3\ldots = .b_1b_2b_3\ldots \leq 1$$

This implies that $.b_1b_2b_3\ldots = 1$ and $.a_1a_2a_3\cdots = 0$. Therefore, every a_k is 0 and, by Lemma 13.6.4, every b_k is 9.

Now suppose that $M = N$ and that the nth decimal place is the first decimal place where the two representations differ; that is, $a_j = b_j$ for all j less than n and a_n is different from b_n. Multiplying by 10^n yields distinct representations

$$a_n.a_{n+1}a_{n+2}a_{n+3}\ldots = b_n.b_{n+1}b_{n+2}b_{n+3}\cdots$$

Since $a_n \neq b_n$, one of them is larger; suppose that a_n is greater than b_n. Then, the first case implies that $a_k = 0$ and $b_k = 9$ for every $k \geq n + 1$. This proves the theorem. $\qquad\square$

One possible way of constructing the real numbers from the integers is to use infinite decimals. However, it is not easy to describe the basic arithmetic operations of addition and multiplication in terms of infinite decimals. Another way of constructing the real numbers is outlined in Problem 15 in Chapter 8.

13.7 Further Examples of Infinite Series

Definition 13.7.1. The *harmonic series* is the series

$$1 + \frac{1}{2} + \frac{1}{3} + \frac{1}{4} + \cdots + \frac{1}{n} + \cdots$$

Theorem 13.7.2. *The harmonic series diverges.*

Proof. This will follow from Theorem 13.4.5 if we show that the set of partial sums does not have an upper bound. We do this by establishing that, for every natural number M, there is a partial sum that is greater than $\frac{1}{2} \cdot M$.

Begin by observing that $\frac{1}{3} + \frac{1}{4}$ is greater than $\frac{1}{4} + \frac{1}{4} = \frac{1}{2}$. Similarly, $\frac{1}{5} + \frac{1}{6} + \frac{1}{7} + \frac{1}{8}$ is greater than $\frac{1}{8} + \frac{1}{8} + \frac{1}{8} + \frac{1}{8} = \frac{1}{2}$. Moreover,

$$\frac{1}{9} + \frac{1}{10} + \cdots + \frac{1}{16}$$

is greater than

$$\frac{1}{16} + \frac{1}{16} + \cdots + \frac{1}{16} = \frac{8}{16} = \frac{1}{2}$$

In general, for every natural number k, each of the terms of the harmonic series from $\frac{1}{2^{k-1}+1}$ to $\frac{1}{2^k}$ is at least $\frac{1}{2^k}$, and there are $2^k - 2^{k-1} = 2^{k-1} \cdot (2 - 1) = 2^{k-1}$ such terms. Therefore, the contribution of the sum of those terms to the harmonic series is at least $2^{k-1} \cdot \frac{1}{2^k} = \frac{1}{2}$.

The above shows that, for every natural number k, the partial sum S_{2k} is at least $\frac{1}{2} \cdot k$. Thus, for every natural number M, there are partial sums of the harmonic series which are greater than $\frac{1}{2} \cdot M$, so the series diverges.   □

Example 13.7.3. The series

$$1 + \frac{1}{\sqrt{2}} + \frac{1}{\sqrt{3}} + \cdots + \frac{1}{\sqrt{n}} + \cdots$$

diverges.

Proof. For every natural number n, $\sqrt{n} \leq n$, so $\frac{1}{\sqrt{n}} \geq \frac{1}{n}$. Therefore, the given series diverges by part (ii) of the Comparison Test (13.5.3), comparing it to the harmonic series (Theorem 13.7.2).   □

Example 13.7.4. The series

$$\frac{1}{1 \cdot 2} + \frac{1}{2 \cdot 3} + \cdots + \frac{1}{n(n+1)} + \cdots$$

converges to 1.

Proof. By Problem 2 in Chapter 2, the nth partial sum of this series is $\frac{n}{n+1}$. It is apparent that this is close to 1 if n is large. To formally establish this, let ε be any positive number. If N is a natural number such that $N + 1 > \frac{1}{\varepsilon}$, then $\varepsilon > \frac{1}{N+1}$. Therefore, if $n \geq N$, then $\left| 1 - \frac{n}{n+1} \right| = \frac{1}{n+1} \leq \frac{1}{N+1} < \varepsilon$.   □

Example 13.7.5. The infinite series

$$1 + \frac{1}{2^2} + \frac{1}{3^2} + \frac{1}{4^2} + \cdots + \frac{1}{n^2} + \cdots$$

converges.

Proof. Since the partial sums of

$$1 + \frac{1}{2^2} + \frac{1}{3^2} + \frac{1}{4^2} + \cdots + \frac{1}{n^2} + \cdots$$

are obtained by adding 1 to those of

$$\frac{1}{2^2} + \frac{1}{3^2} + \cdots + \frac{1}{n^2} + \cdots$$

it suffices to show that this latter series converges. We will establish this by comparison to the convergent series from Example 13.7.4. The series

$$\frac{1}{2^2} + \frac{1}{3^2} + \cdots + \frac{1}{(n+1)^2} + \cdots$$

is term by term less than the series

$$\frac{1}{1 \cdot 2} + \frac{1}{2 \cdot 3} + \cdots + \frac{1}{n(n+1)} + \cdots$$

since $\frac{1}{(n+1)^2}$ is less than $\frac{1}{n(n+1)}$ for every natural number n. Therefore, the series

$$\frac{1}{2^2} + \frac{1}{3^2} + \cdots + \frac{1}{(n+1)^2} + \cdots$$

converges by the Comparison Test (13.5.3).   □

It can be shown, using calculus, that the series in the above example converges to $\frac{\pi^2}{6}$.

Theorem 13.7.6. *For every $p \geq 2$, the series*

$$1 + \frac{1}{2^p} + \frac{1}{3^p} + \cdots + \frac{1}{n^p} + \cdots$$

converges.

Proof. This follows immediately from the previous example and the fact that $p \geq 2$ implies that $\frac{1}{n^p} \leq \frac{1}{n^2}$ for every natural number n.   □

It can be proven (using integral calculus) that, in fact, $1 + \frac{1}{2^p} + \frac{1}{3^p} + \cdots$ converges for every $p > 1$.

Infinite series are often used to define specific numbers. For example, the famous number e, the base of the natural logarithm, can be defined as an infinite series. (It can also be defined in many other ways.)

Example 13.7.7. The series $1 + \frac{1}{1!} + \frac{1}{2!} + \frac{1}{3!} + \cdots + \frac{1}{n!} + \cdots$ converges. The sum of the series is denoted by e.

Proof. An easy application of the Generalized Principle of Mathematical Induction shows that $n! > 2^n$ for all $n \geq 4$ (see Theorem 2.1.5 of Chapter 2). Thus, $\frac{1}{n!} < \frac{1}{2^n}$ for all $n \geq 4$. Hence, $\frac{1}{4!} + \frac{1}{5!} + \frac{1}{6!} + \cdots$ converges by the Comparison Test (13.5.3), and therefore so does the entire series.   □

We next consider a particularly interesting example of a divergent series.

Theorem 13.7.8. *Let p_j denote the jth prime number (so that $p_1 = 2$, $p_2 = 3$, $p_3 = 5$, etc.). Then the series*

$$\frac{1}{2} + \frac{1}{3} + \frac{1}{5} + \cdots + \frac{1}{p_j} + \cdots$$

diverges. That is, the series of the reciprocals of the primes diverges.

Proof. This can be proven in several ways. The proof that we present below appears to be the easiest, although it is somewhat tricky. We show that the assumption that the series converges leads to a contradiction.

Suppose that the sum of the series was S. Then, by the definition of convergence (Definition 13.1.4), there would be a partial sum of the series, say S_k, such that S_k is within $\frac{1}{2}$ of S. It would then follow that the sum of the series obtained by discarding the first k terms would be less than $\frac{1}{2}$. That is, there would be a natural number k such that

$$\frac{1}{p_{k+1}} + \frac{1}{p_{k+2}} + \frac{1}{p_{k+3}} + \cdots$$

is less than $\frac{1}{2}$. We proceed to show that this is impossible.

For the rest of the proof, we fix such a k and say that p_j is a "small prime" if j is less than or equal to k and that p_j is a "big prime" if j is greater than k. For each natural number x, let $N(x)$ denote the number of natural numbers that are less than or equal to x and are not divisible by any big prime. The surprising trick involves obtaining, and using, an upper bound on $N(x)$.

Fix any natural number x. We can get a crude upper bound for $N(x)$ as follows. Every natural number y that is counted in $N(x)$ can be written as a product uv, where u is a perfect square and v does not have any perfect square divisors. To see this, use the prime factorization (Corollary 4.1.2) of y to factor out the biggest perfect square u; v is the other factor of y. Note that $u = 1$ if 1 is the largest perfect square that divides y, and $v = 1$ if y itself is a perfect square. The number of distinct u's that arise from y's that are counted in $N(x)$ must be less than or equal to $\sqrt{x}$, since each u is less than or equal to x and is therefore the square of a number that is less than or equal to $\sqrt{x}$. Also, each of the v's consists of a product of small primes raised to at most the first power. There are k small primes, so the number of possible v's is at most 2^k (since each small prime may or may not occur in the prime factorization of each v). Every y that is counted by $N(x)$ is of the form uv and there are at most $\sqrt{x}$ u's and 2^k v's, from which it follows that $N(x) \le 2^k \sqrt{x}$, for each x. We will use this inequality to derive a contradiction.

Next, we establish a lower bound for $N(x)$ that will be inconsistent with the above inequality when x is large. First, for any prime p and any natural number x, there are at most $\frac{x}{p}$ multiples of p that are less than or equal to x. Thus there are at most $\frac{x}{p_j}$ natural numbers less than or equal to x that have the big prime p_j as a factor. Remember that $N(x)$ denotes the number of natural numbers less than or equal to x that have only small prime factors. Therefore, $x - N(x)$ is the number of natural numbers less than or equal to x having at least one big prime as a factor. There are at most $\frac{x}{p_{k+1}}$ of those that have the big prime p_{k+1} as a factor; there are at most $\frac{x}{p_{k+2}}$ of those that have the big prime p_{k+2} as a factor; and so on. Thus, $x - N(x)$ is less than or equal to $\frac{x}{p_{k+1}} + \frac{x}{p_{k+2}} + \frac{x}{p_{k+3}} + \cdots$. Each partial sum of this series is x times a partial sum of the series $\frac{1}{p_{k+1}} + \frac{1}{p_{k+2}} + \frac{1}{p_{k+3}} + \cdots$. Since the sum of the latter series is less than $\frac{1}{2}$, it follows that $x - N(x)$ is less than $\frac{x}{2}$.

The inequality $x - N(x) < \frac{x}{2}$ is equivalent to $\frac{x}{2} < N(x)$. Combining this with the previous inequality we obtained for $N(x)$ gives

$$\frac{x}{2} < N(x) \le 2^k \sqrt{x}$$

Therefore,

$$\frac{x}{2} < 2^k \sqrt{x}$$

Multiplying both sides of this inequality by 2 and dividing by $\sqrt{x}$ yields

$$\sqrt{x} < 2^{k+1}$$

Thus, assuming that the series converges leads to the conclusion that there is a natural number k such that $\sqrt{x} < 2^{k+1}$ for every natural number x. Now we have our contradiction: Since k is fixed, the above cannot hold for all natural numbers x. For example, if $x = 2^{2k+4}$, then $\sqrt{x} = 2^{k+2}$, which is larger than 2^{k+1}. □

The preceding provides another proof that there are infinitely many primes; if there were only a finite number of primes, the series consisting of the reciprocals of the primes would have only a finite number of terms, and therefore would converge.

In Chapter 1, we mentioned the famous unsolved twin primes problem. While (in spite of dramatic progress by Yitang Zhang and others, beginning in 2013) the twin primes problem is still unsolved, Viggo Brun proved in 1919 that the sum of the reciprocals of the twin primes converges. (The proof of this is well beyond the scope of this book.)

There is much more that is known about infinite series. Some other results are outlined in the interesting and challenging problems below. Almost every calculus book contains a large amount of related material.

13.8 Problems

Basic Exercises

1. Find the sum of each of the following geometric series:
   (a) $1 - \frac{1}{3} + \frac{1}{3^2} - \frac{1}{3^3} + \cdots$
   (b) $10 + \frac{10}{7} + \frac{10}{7^2} + \frac{10}{7^3} + \cdots$
   (c) $\frac{3}{\sqrt{2}} + \frac{3}{\left(\sqrt{2}\right)^2} + \frac{3}{\left(\sqrt{2}\right)^3} + \frac{3}{\left(\sqrt{2}\right)^4} + \cdots$

2. For each of the following sets, determine if the set has an upper bound, and, if so, find the least upper bound:

   (a) $\{1, -1, 2, -2, 3, -3, \dots\}$
   (b) $\{1, -1, \frac{1}{2}, -\frac{1}{2}, \frac{1}{3}, -\frac{1}{3}, \dots\}$
   (c) $\{-1, -\frac{1}{2}, -\frac{1}{3}, -\frac{1}{4}, \dots\}$
   (d) $\{-7, 12, -4, 6\}$

3. Determine which of the following series converges:

   (a) $1 - \frac{1}{7} + \frac{1}{7^2} - \frac{1}{7^3} + \frac{1}{7^4} - \cdots$
   (b) $\sqrt{2} + 789 + 2042 + \frac{3}{2} + \frac{3}{2^2} + \frac{3}{2^3} + \cdots + \frac{3}{2^n} + \cdots$
   (c) $.1 + .1 + .1 + .1 + \cdots$
   (d) $1 + \frac{1}{2} + \frac{1}{3} + \frac{1}{2} + \frac{1}{4} + \frac{1}{2} + \frac{1}{5} + \frac{1}{2} + \cdots$
   (e) $1 + \frac{1}{2} + \frac{1}{2^2} + \frac{1}{2} + \frac{1}{2^3} + \frac{1}{2} + \frac{1}{2^4} + \frac{1}{2} + \cdots$
   (f) $\frac{5}{3^2} + \frac{17}{3^3} + \frac{5}{3^4} + \frac{17}{3^5} + \cdots$

4. For each of the following infinite decimals, determine the rational number that it represents:

   (a) $.77777777\dots$
   (b) $.3434343434\dots$
   (c) $17.389389389389\dots$

5. Let t be any positive real number. Use the Principle of Mathematical Induction to prove that $(1+t)^n \geq 1 + nt$ for every natural number n. (This result was used without proof in Theorem 13.2.3.)

6. Determine which of the following series converge:

   (a) $\frac{1}{2} + \frac{2}{3} + \frac{3}{4} + \frac{4}{5} + \frac{5}{6} + \cdots + \frac{n}{n+1} + \cdots$
   (b) $\frac{1}{5} + \frac{1}{9} + \frac{1}{17} + \cdots + \frac{1}{2^n+1} + \cdots$

Interesting Problems

7. Determine which of the following series converge:

   (a) $19 + \frac{19}{2^{7/2}} + \frac{19}{3^{7/2}} + \frac{19}{4^{7/2}} + \cdots + \frac{19}{n^{7/2}} + \cdots$
   (b) $\frac{1}{2} + \frac{1}{3} + \frac{1}{2^2} + \frac{1}{3^2} + \frac{1}{2^3} + \frac{1}{3^3} + \cdots + \frac{1}{2^n} + \frac{1}{3^n} + \cdots$

8. Determine the rational number that is represented by each of the following infinite decimals:

   (a) $6.798345345345345345\dots$
   (b) $-38.0006561234123412341234\dots$
   (c) $.012345678901234567890123456789\dots$

9. Suppose that the sum of the series $a_1 + a_2 + a_3 + \cdots + a_n + \cdots$ is S and the sum of the series $b_1 + b_2 + b_3 + \cdots + b_n + \cdots$ is T. Prove that the sum of the series

$$(a_1 - b_1) + (a_2 - b_2) + (a_3 - b_3) + \cdots + (a_n - b_n) + \cdots$$

is $S - T$.

10. ("Sigma notation") There is a standard notation that is often used when considering infinite series and in many other contexts. The Greek letter $\sum$ (called "sigma") is part of a shorthand for representing sums, as illustrated by the following examples:

(i) $\sum_{i=1}^{4} a_i$ is defined to be $a_1 + a_2 + a_3 + a_4$; it can be read "the sum from $i = 1$ to 4 of a_i"

(ii) $\sum_{n=3}^{5} \frac{n}{2}$ means $\frac{3}{2} + \frac{4}{2} + \frac{5}{2}$; it can be read "the sum from $n = 3$ to 5 of $\frac{n}{2}$"

(iii) $\sum_{j=5}^{21} j^2$ means $5^2 + 6^2 + 7^2 + \cdots + 21^2$; it can be read "the sum from $j = 5$ to 21 of j^2"

(iv) $\sum_{i=1}^{\infty} \frac{1}{2^i}$ means $\frac{1}{2} + \frac{1}{2^2} + \frac{1}{2^3} + \cdots + \frac{1}{2^i} + \cdots$; it can be read "the sum from $i = 1$ to infinity of $\frac{1}{2^i}$"

Thus, $\sum$ is used as above to indicate sums. When we write $\sum_{i=1}^{\infty} a_i$ we do not necessarily imply that the series converges; it is merely shorthand for the infinite series $a_1 + a_2 + a_3 + \cdots$. If the series converges to S, we may write $\sum_{i=1}^{\infty} a_i = S$. For example, $\sum_{i=1}^{\infty} \frac{1}{2^i} = 1$.

(a) Find: $\sum_{i=1}^{4} \frac{1}{i}$

(b) Find: $\sum_{i=3}^{7} (-1)^i \cdot (i + 4)$

(c) Find: $\sum_{i=1}^{\infty} \frac{7}{4^i}$

(d) Find: $\sum_{i=5}^{\infty} (-1)^i \cdot \frac{19}{5^i}$

(e) Find: $\sum_{i=1}^{100} (-1)^i$

(f) Find: $\sum_{i=2}^{17} \left(\frac{1}{i} - \frac{1}{i+1} \right)$

(g) Show that $m \sum_{i=1}^{n} a_i = \sum_{i=1}^{n} ma_i$, where n is any natural number and m is any real number.

(h) Show that $\left(\sum_{i=1}^{n} a_i \right) + \left(\sum_{i=1}^{n} b_i \right) = \sum_{i=1}^{n} (a_i + b_i)$.

11. Show that there is no least upper bound property for the set of rational numbers. In other words, show that there are nonempty sets of rational numbers that have rational upper bounds but which do not have a least rational upper bound.

Challenging Problems

12. Show that the series $1 + \frac{1}{3} + \frac{1}{5} + \frac{1}{7} + \cdots$ diverges.

13. (a) Suppose that the series $a_1 + a_2 + a_3 + \cdots$ converges. Prove that, for each positive number δ, there is a natural number N such that $|a_i| \leq \delta$ for every $i \geq N$.

    (b) Suppose that there is a positive number δ such that $|a_i| \geq \delta$ for infinitely many a_i. Show that the series $a_1 + a_2 + a_3 + \cdots$ diverges.

    (c) Let $a_i = (-1)^i \frac{i}{i+100}$. Show that the series $a_1 + a_2 + a_3 + \cdots$ diverges.

14. (A form of the "Ratio Test")

    (a) Show that the series $a_1 + a_2 + a_3 + \cdots$ with nonnegative terms converges if there is a positive number $r < 1$ such that a_{i+1} is less than or equal to ra_i for every i.

    [Hint: Compare the given series to the geometric series $a_1 + a_1 r + a_1 r^2 + a_1 r^3 + \cdots$.]

    (b) Show that the series $a_1 + a_2 + a_3 + \cdots$ with nonnegative terms converges if there exist an N and a positive number $r < 1$ such that a_{i+1} is less than or equal to ra_i for every $i \geq N$.

15. (Absolute convergence implies convergence) The series $a_1 + a_2 + a_3 + \cdots$ is said to *converge absolutely* if the series $|a_1| + |a_2| + |a_3| + \cdots$ converges. The following is an outline of a proof that a series converges if it converges absolutely.

    Suppose that $|a_1| + |a_2| + |a_3| + \cdots$ converges.

    (a) Show that $2|a_1| + 2|a_2| + 2|a_3| + \cdots$ converges.

    (b) Use the Comparison Test (13.5.3) to show that $(|a_1| + a_1) + (|a_2| + a_2) + (|a_3| + a_3) + \cdots$ converges.

    (c) Prove that the series $-|a_1| - |a_2| - |a_3| - \cdots$ converges, and add it to the series in part (b) to show that $a_1 + a_2 + a_3 + \cdots$ converges.

16. Prove that, for every real number x, the series $1 + x + \frac{x^2}{2!} + \frac{x^3}{3!} + \frac{x^4}{4!} + \cdots$ converges. (This is one definition of the exponential function e^x, where e is the base of the natural logarithm. That is, $e^x = 1 + x + \frac{x^2}{2!} + \frac{x^3}{3!} + \frac{x^4}{4!} + \cdots$.)

[Hint: Since absolute convergence implies convergence (Problem 15), it suffices to prove convergence for positive x. For this, use the Ratio Test (Problem 14) with N any natural number larger than x and with $r = \frac{x}{x+1}$.]

17. (A form of the "Root Test") Suppose that $a_1 + a_2 + a_3 + \cdots$ is a series with nonnegative terms. Prove that the series converges if there is a positive number $r < 1$ and a natural number N such that $(a_i)^{\frac{1}{i}} \le r$ for all $i \ge N$.

18. The *alternating harmonic series* is the series:

$$1 - \frac{1}{2} + \frac{1}{3} - \frac{1}{4} + \cdots + \frac{(-1)^{n+1}}{n} + \cdots$$

(a) Let $\mathcal{T}$ denote the set of even partial sums of the alternating harmonic series; that is, $\mathcal{T} = \{1 - \frac{1}{2}, 1 - \frac{1}{2} + \frac{1}{3} - \frac{1}{4}, 1 - \frac{1}{2} + \frac{1}{3} - \frac{1}{4} + \frac{1}{5} - \frac{1}{6}, \ldots\}$. Show that 1 is an upper bound for $\mathcal{T}$.

(b) Let S be the least upper bound of $\mathcal{T}$. Show that the alternating harmonic series converges to S.

19. (a) Show that the terms of the alternating harmonic series $1 - \frac{1}{2} + \frac{1}{3} - \frac{1}{4} + \cdots$ can be rearranged so that the resulting series converges to 7.
   [Hint: Use the divergence of the series $1 + \frac{1}{3} + \frac{1}{5} + \cdots$ (Problem 12) to get the first partial sum that is greater than 7. Begin the rearrangement with that partial sum. Then, adding the first negative term, $-\frac{1}{2}$, will make the sum less than 7. Then add positive terms until the result is just greater than 7; then add negative terms to get less than 7, and so on.]

(b) Let t be any real number. Show that the terms of the alternating harmonic series can be rearranged so that the resulting series converges to t.

20. Give an example of a series which converges but does not converge absolutely (see Problem 15).

21. A series is said to *converge conditionally* if it converges but does not converge absolutely. Prove that any conditionally convergent series can be rearranged to sum to any real number, and can also be rearranged so that it diverges.
   [Hint: First show that the sum of the nonnegative terms of the series diverges, as does the sum of the negative terms of the series.]

22. Prove that, if a series converges absolutely, then all the rearrangements of the series have the same sum.

23. (Characterization of the rational numbers) A *repeating infinite decimal* is an infinite decimal of the form:

$$L.a_1 a_2 \cdots a_m b_1 b_2 \cdots b_n b_1 b_2 \cdots b_n b_1 b_2 \cdots b_n \cdots$$

where L is an integer and the a_i and b_i are digits.

(a) Show that every repeating infinite decimal represents a rational number.

(b) Show that every rational number has a representation as a repeating infinite decimal.

24. A number t is said to be a *lower bound* for a set S of real numbers if t is less than or equal to x for every x in S. A number t_0 is a *greatest lower bound* for the set S if t_0 is a lower bound for S and t_0 is greater than or equal to t for every lower bound t of S. Prove that every nonempty set of real numbers that has a lower bound has a greatest lower bound.
    [Hint: Consider the set $T = \{-x : x \in S\}$.]

25. (a) Prove that, if a series converges, then the set of all its partial sums has a lower bound (see Problem 24).
    (b) Prove that a series whose terms are all less than or equal to 0 converges if and only if the set of all its partial sums has a lower bound.

26. (Construction of least upper bounds) For this problem we assume familiarity with the construction of the real numbers from sets of rational numbers using Dedekind cuts, as outlined in Problem 15 in Chapter 8. In that context, the real number A is said to be *less than or equal to* the real number B if A is contained in B.

    Suppose that a given nonempty set S of real numbers (which is a set of sets of rational numbers) has an upper bound. That is, there is a real number that is greater than or equal to every real number in S. Prove that the union of all of the real numbers in S is a real number and is the least upper bound of S.

27. A more standard approach to convergence of infinite series begins with the definition of convergence of any sequence of real numbers. The definition of convergence that we have given (13.1.4) is the particular case when the sequence is the sequence of partial sums of a series. We have delayed the presentation of the more general definition of convergence of sequences because some people find it more confusing to learn the general definition when they are first exposed to this topic. However, the general definition is required in order to obtain many of the standard results on infinite series, and in many other situations. In this exercise, we state the definition of convergence of an arbitrary sequence of real numbers, and use that definition to reformulate and extend some of the results obtained in the chapter and in the problems given above.

    Some examples of sequences are:

    $$1, \frac{1}{2}, \frac{1}{3}, \frac{1}{4}, \ldots$$

    $$\sqrt{3}, \sqrt{4}, \sqrt{5}, \ldots$$

    $$1, -1, 1, -1, 1, -1, \ldots$$

    In general, a *sequence* of real numbers is a listing of real numbers, one for each natural number. More precisely, a sequence of real numbers can be defined as an assignment of a real number to each natural number (that is, as a function from the natural numbers to the real numbers). For example, the sequence

$1, \frac{1}{2}, \frac{1}{3}, \frac{1}{4}, \ldots$ is the assignment of the number $\frac{1}{n}$ to each natural number n, and the sequence $1, -1, 1, -1, 1, -1, \ldots$ is the assignment of 1 to every odd natural number and -1 to every even natural number.

Notation such as $x_1, x_2, x_3, x_4, \ldots$ is often used to denote a sequence. Sometimes we abbreviate this as (x_n). The crucial concept is that of a limit of a sequence. The rough idea is that a real number L is the limit of the sequence (x_n) if x_n is close to L when n is large. More accurately, no matter how close we wish to get x_n to L, it will be that close if n is sufficiently large. The precise definition is the following.

Definition 13.8.1. The sequence (x_n) *converges to L* (or *has limit L*) if, for every real number ε greater than 0, there is a natural number N such that $|x_n - L| < \varepsilon$ whenever $n \geq N$. We use the notation $\lim\limits_{n \to \infty} x_n = L$ to denote the fact that L is the limit of the sequence (x_n); this is read "the limit as n approaches infinity of the sequence (x_n) is L." Sometimes the notation $(x_n) \to L$ is used.

Note that applying this definition to a sequence (S_n) of partial sums of an infinite series yields Definition 13.1.4. That is, an infinite series converges to S if and only if the sequence of its partial sums converges to S. In studying an infinite series $a_1 + a_2 + a_3 + \cdots$, there are several different sequences that naturally arise. The one that we have discussed so far is the sequence (S_n) of partial sums. Another is the sequence (a_i) of terms of the series. It is important not to confuse the two. Limits of some other sequences, including some related to the terms of a series, also play a role.

(a) Prove that, if a series converges, then its sequence of terms converges to 0 (see Problem 13). That is, if $a_1 + a_2 + a_3 + \cdots$ converges, then $\lim\limits_{i \to \infty} a_i = 0$.

(b) ("The Ratio Test") Suppose that $a_1 + a_2 + a_3 + \cdots$ is a series of non-zero terms and suppose that $\lim\limits_{i \to \infty} \left| \dfrac{a_{i+1}}{a_i} \right| = r$.

   (i) Show that the series $a_1 + a_2 + a_3 + \cdots$ converges absolutely if $r < 1$. [Hint: See Problems 14 and 15.]
   (ii) Show that the series $a_1 + a_2 + a_3 + \cdots$ diverges if $r > 1$.
   (iii) Give an example of a series that diverges and has $r = 1$.
   (iv) Give an example of a series that converges and has $r = 1$.

(c) ("The Root Test") Suppose that $\lim\limits_{i \to \infty} |a_i|^{\frac{1}{i}} = r$.

   (i) Show that the series $a_1 + a_2 + a_3 + \cdots$ converges absolutely if $r < 1$. [Hint: See Problems 15 and 17.]
   (ii) Show that the series $a_1 + a_2 + a_3 + \cdots$ diverges if $r > 1$.

(d) ("The Limit Comparison Test") Let $a_1 + a_2 + a_3 + \cdots$ and $b_1 + b_2 + b_3 + \cdots$ be series whose terms are all positive. Suppose that $\lim\limits_{i \to \infty} \dfrac{a_i}{b_i} = r$ for some

$r > 0$. Show that the series $a_1 + a_2 + a_3 + \cdots$ converges if and only if the series $b_1 + b_2 + b_3 + \cdots$ converges.

(e) Determine which of the following series converge.

[Hint: It may be useful to use some of the results from this chapter, as well as some of the previous parts of this problem.]

(i) $-2 + \frac{2}{3} - \frac{2}{4} + \frac{2}{5} - \frac{2}{6} + \cdots$

(ii) $\frac{1}{\sqrt{11}} + \frac{2}{\sqrt{11^2}} + \frac{3}{\sqrt{11^3}} + \cdots + \frac{n}{\sqrt{11^n}} + \cdots$

(iii) $\frac{1^{100}}{3} + \frac{2^{100}}{3^2} + \frac{3^{100}}{3^3} + \frac{4^{100}}{3^4} + \cdots + \frac{n^{100}}{3^n} + \cdots$

(iv) $\frac{1}{1+1} + \frac{1}{2+\frac{1}{2}} + \frac{1}{3+\frac{1}{3}} + \cdots + \frac{1}{n+\frac{1}{n}} + \cdots$

(v) $\frac{1}{7^2-3\cdot7} + \frac{1}{8^2-3\cdot8} + \frac{1}{9^2-3\cdot9} + \cdots + \frac{1}{n^2-3n} + \cdots$

Chapter 14
Some Higher Dimensional Spaces

Spaces of two and three dimensions are familiar to most people. Four-dimensional space, however, seems more mysterious. Nonetheless, spaces of dimension four and higher have been defined by mathematicians in ways that are easy to understand. In this chapter, we describe some spaces of various dimensions, including infinite-dimensional spaces.

14.1 Two-Dimensional Space

The plane, the basic two-dimensional space, can be represented using coordinates. We consider two mutually perpendicular axes in the plane, one horizontal and one vertical, and represent points in the plane as pairs of numbers relative to those axes. Instead of calling those axes the x and y axes, as is most common, in this chapter we prefer to call them the horizontal and vertical axes. The point of intersection of the axes is called the *origin*.

Each point in the plane is assigned an ordered pair of real numbers as coordinates. A point is assigned the coordinates (x_1, x_2) as follows. The first coordinate, x_1, is the distance from the point to the vertical axis if the point is to the right of the vertical axis and is the negative of the distance from the point to the vertical axis if the point is to the left of the vertical axis. If the point is on the vertical axis, then $x_1 = 0$. Similarly, the point has second coordinate x_2 equal to the distance from the point to the horizontal axis if the point is above the horizontal axis and the negative of its distance to the horizontal axis if the point is below the horizontal axis. If the point is on the horizontal axis, then $x_2 = 0$.

Some examples of coordinates of points are illustrated in Figure 14.1.

The above associates a unique ordered pair of real numbers to each point in the plane. Conversely, if (x_1, x_2) is any pair of real numbers, then (x_1, x_2) corresponds to a unique point in the plane.

© Springer Nature Switzerland AG 2018
D. Rosenthal et al., *A Readable Introduction to Real Mathematics*,
Undergraduate Texts in Mathematics, https://doi.org/10.1007/978-3-030-00632-7_14

Fig. 14.1 Some points in the plane

The geometry of the plane can be expressed in terms of coordinates. For example, one fundamental concept is that of the distance between two points. The distance formula is the following.

Theorem 14.1.1. *If $x = (x_1, x_2)$ and $y = (y_1, y_2)$ are points in the plane, then the distance from x to y is*

$$\sqrt{(x_1 - y_1)^2 + (x_2 - y_2)^2}$$

Proof. We first consider some special cases. If $x_1 = y_1$, then the two points lie on the same vertical line and so the distance between them is $|x_2 - y_2|$. Since $x_1 - y_1 = 0$, this is equal to $\sqrt{(x_1 - y_1)^2 + (x_2 - y_2)^2}$. Similarly, if $x_2 = y_2$, then the two points lie on the same horizontal line and the result follows.

Now assume that $x_1 \neq y_1$ and $x_2 \neq y_2$. Consider the triangle whose vertices are (x_1, x_2), (y_1, y_2), and (x_1, y_2). We illustrate the situation in the case where all the coordinates are positive in Figure 14.2, but the proof is the same in all cases. Note that the triangle is a right triangle.

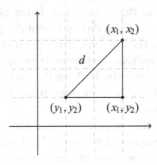

Fig. 14.2 Proof of the distance formula

The distance between the points (x_1, x_2) and (y_1, y_2) is the length of the hypotenuse of the triangle; let's call this distance d. The legs have lengths $|x_1 - y_1|$ and $|x_2 - y_2|$. By the Pythagorean Theorem (11.3.6), $d^2 = |x_1 - y_1|^2 + |x_2 - y_2|^2$. Since $|x_1 - y_1|^2 = (x_1 - y_1)^2$ and $|x_2 - y_2|^2 = (x_2 - y_2)^2$, it follows that

$$d^2 = (x_1 - y_1)^2 + (x_2 - y_2)^2$$

Thus, $d = \sqrt{(x_1 - y_1)^2 + (x_2 - y_2)^2}$. $\qquad\qquad\square$

It is often useful to consider the points in the plane as being represented by line segments emanating from the origin. Such line segments are called *vectors*. For example, the vector $(-1, 2)$ is represented by the line segment from the origin to the point $(-1, 2)$ (see Figure 14.3). The *zero vector* is simply $(0, 0)$.

Fig. 14.3 A vector in the plane

Another fundamental geometric concept is that of an angle between two vectors. The angle between the vectors (x_1, x_2) and (y_1, y_2) can be expressed in terms of their coordinates (see Theorem 14.4.3). However, for present purposes, we only consider the characterization of perpendicularity.

Theorem 14.1.2. *The vectors (x_1, x_2) and (y_1, y_2) are perpendicular to each other if and only if $x_1 y_1 + x_2 y_2 = 0$.*

Proof. Notice that if one, or both, of the two vectors is the zero vector, then $x_1 y_1 + x_2 y_2 = 0$. For this reason we define the zero vector as being perpendicular to every vector. Another special case is when the two vectors lie on the same line through the origin; that is, when $y_1 = tx_1$ and $y_2 = tx_2$ for some real number t. In this case, the vectors are perpendicular if and only if one of them is the zero vector, which happens if and only if $tx_1 = tx_2 = 0$. Thus, the theorem holds in this case.

Now assume that the two vectors do not lie on the same line through the origin. Consider the triangle with vertices (x_1, x_2), (y_1, y_2), and the origin, $(0, 0)$, as pictured in Figure 14.4. The vector (x_1, x_2) is perpendicular to the vector (y_1, y_2) if and only if the angle of this triangle at the vertex $(0, 0)$ is 90 degrees. By the Pythagorean Theorem (11.3.6) and its converse (Problem 15 in Chapter 11), this holds if and only if the sum of the squares of the lengths of the sides from $(0, 0)$ to (x_1, x_2) and from $(0, 0)$ to (y_1, y_2) is equal to the square of the distance between

Fig. 14.4 Perpendicularity of vectors

(x_1, x_2) and (y_1, y_2). The squares of those lengths are, respectively, $x_1^2 + x_2^2$, $y_1^2 + y_2^2$, and $(x_1 - y_1)^2 + (x_2 - y_2)^2$. The latter is equal to

$$x_1^2 - 2x_1 y_1 + y_1^2 + x_2^2 - 2x_2 y_2 + y_2^2$$

Thus, the vectors are perpendicular if and only if

$$x_1^2 + x_2^2 + y_1^2 + y_2^2 = x_1^2 - 2x_1 y_1 + y_1^2 + x_2^2 - 2x_2 y_2 + y_2^2$$

This equality holds if and only if $-2x_1 y_1 - 2x_2 y_2 = 0$, which is equivalent to $x_1 y_1 + x_2 y_2 = 0$.   □

The plane with its standard geometry is called *two-dimensional Euclidean space*. Because of its representation as pairs of real numbers, it is often denoted $\mathbb{R}^2$.

14.2   Three-Dimensional Space

The space that we live in appears to be three-dimensional. Locating points in three-dimensional space requires a triple of numbers. Start with a plane in which there are mutually perpendicular horizontal and vertical axes, as discussed above. Each point in three-dimensional space is either on, above, or below the given plane. We assign a triple of coordinates to each point, as follows. Introduce a third axis perpendicular to the plane and going through the intersection of the horizontal and vertical axes; the horizontal, vertical, and third axes are mutually perpendicular. The point of intersection of these three axes is called the *origin*.

To assign a triple of real numbers to each point, begin by dropping a perpendicular from the point to the plane. The perpendicular intersects the plane in some point; let (x_1, x_2) be the coordinates of that point in the plane. Now let x_3 be the length of that perpendicular to the plane if the point is above the plane, 0 if the point is on the plane, and the negative of the length of that perpendicular if the point is below the plane. The triple (x_1, x_2, x_3) gives the coordinates of the point.

As in the two-dimensional case, the geometry of three-dimensional space can be captured in terms of the coordinates of points. There is a distance formula that is similar to the formula in two-dimensional space.

Theorem 14.2.1. *The distance between the points (x_1, x_2, x_3) and (y_1, y_2, y_3) is*

$$\sqrt{(x_1 - y_1)^2 + (x_2 - y_2)^2 + (x_3 - y_3)^2}$$

Proof. Begin by considering the case where the two points agree in one of their coordinates. We prove the case where $x_3 = y_3$. (The proof is similar if $x_1 = y_1$ or $x_2 = y_2$.) Then the two points both lie in a plane which is either on (if $x_3 = 0$), above (if x_3 is greater than 0), or below (if x_3 is less than 0) the plane consisting of all points whose third coordinate is 0. The distance between the points is the same as the distance between (x_1, x_2) and (y_1, y_2), which is $\sqrt{(x_1 - y_1)^2 + (x_2 - y_2)^2}$ by Theorem 14.1.1. Since $x_3 = y_3$, $(x_3 - y_3)^2 = 0$, so the distance between the points is $\sqrt{(x_1 - y_1)^2 + (x_2 - y_2)^2 + (x_3 - y_3)^2}$.

The general case can be obtained from the above special case, as follows. Consider the triangle with vertices (x_1, x_2, x_3), (y_1, y_2, y_3) and (y_1, y_2, x_3), as in Figure 14.5. (In the diagram, the case where y_3 is greater than x_3 is shown; if y_3 is less than x_3, the picture is similar.)

Fig. 14.5 Proving the distance formula in three-dimensional space

This is a right triangle since the line segment from (x_1, x_2, x_3) to (y_1, y_2, x_3) lies in the plane consisting of all points whose third coordinate is x_3, and the point (y_1, y_2, y_3) is directly above or below the point (y_1, y_2, x_3). The length of the hypotenuse of this right triangle is the distance from (x_1, x_2, x_3) to (y_1, y_2, y_3). By the special case, the square of the distance from (x_1, x_2, x_3) to (y_1, y_2, x_3) is $(x_1 - y_1)^2 + (x_2 - y_2)^2$. The square of the distance between the points (y_1, y_2, y_3) and (y_1, y_2, x_3) is $(x_3 - y_3)^2$, since one of those points lies directly above the other (depending upon which of x_3 and y_3 is larger). By the Pythagorean Theorem (11.3.6), the square of the length of the hypotenuse is $(x_1 - y_1)^2 + (x_2 - y_2)^2 + (x_3 - y_3)^2$, which proves the formula. □

As in the two-dimensional case, we frequently think of points in three-dimensional space as corresponding to line segments from the origin to the points, which we call *vectors*. There is an important characterization of perpendicularity of vectors in terms of their coordinates; it is similar to the formula in two dimensions. As in the two-dimensional case, we call the vector whose coordinates are all zero the *zero vector* and regard it as being perpendicular to every vector.

Theorem 14.2.2. *The vectors* (x_1, x_2, x_3) *and* (y_1, y_2, y_3) *are perpendicular to each other if and only if* $x_1 y_1 + x_2 y_2 + x_3 y_3 = 0$.

Proof. As in the two-dimensional case, the result is clearly true if the vectors are multiples of each other. So assume that the vectors are not multiples of each other and consider the triangle with vertices $(0, 0, 0)$, (x_1, x_2, x_3) and (y_1, y_2, y_3). By the Pythagorean Theorem (11.3.6) and its converse (Problem 15 in Chapter 11), the vectors are perpendicular if and only if the sum of the squares of the lengths of the segments from $(0, 0, 0)$ to (x_1, x_2, x_3) and from $(0, 0, 0)$ to (y_1, y_2, y_3) is equal to the square of the distance from (x_1, x_2, x_3) to (y_1, y_2, y_3). Computing the sum of the squares of the lengths of the first two segments gives $x_1^2 + x_2^2 + x_3^2 + y_1^2 + y_2^2 + y_3^2$. The square of the third has length $(x_1 - y_1)^2 + (x_2 - y_2)^2 + (x_3 - y_3)^2$. The latter equals

$$x_1^2 - 2x_1 y_1 + y_1^2 + x_2^2 - 2x_2 y_2 + y_2^2 + x_3^2 - 2x_3 y_3 + y_3^2$$

This is equal to $x_1^2 + x_2^2 + x_3^2 + y_1^2 + y_2^2 + y_3^2$ if and only if $-2x_1 y_1 - 2x_2 y_2 - 2x_3 y_3 = 0$, which is equivalent to $x_1 y_1 + x_2 y_2 + x_3 y_3 = 0$.   □

This space, consisting of triples of real numbers with the distance between two triples given by the distance formula in Theorem 14.2.1, is called *three-dimensional Euclidean space*. Because of its representation as triples of real numbers, it is often denoted $\mathbb{R}^3$.

14.3   Spaces of Dimension Four and Higher

Many people have difficulty with the idea of a four-dimensional space, since they cannot conceive of four axes each of which is perpendicular to all of the other three. It is true that four such axes cannot be constructed within the three-dimensional space that we appear to live in. On the other hand, we can easily think of four-tuples of numbers. In the cases of two and three-dimensional spaces, we started with an understanding of the geometry and represented it in terms of coordinates. For spaces of dimension four (and higher) we reverse the process. We define four-dimensional space in terms of four-tuples, and then introduce geometric ideas using the coordinates.

Definition 14.3.1. *Four-dimensional Euclidean space*, denoted $\mathbb{R}^4$, is the set of all four-tuples of real numbers (x_1, x_2, x_3, x_4), with the distance between the four-tuples (x_1, x_2, x_3, x_4) and (y_1, y_2, y_3, y_4) defined to be

$$\sqrt{(x_1 - y_1)^2 + (x_2 - y_2)^2 + (x_3 - y_3)^2 + (x_4 - y_4)^2}$$

Even without "seeing" four dimensions, we can study four-dimensional space in terms of the coordinates of points. As in $\mathbb{R}^2$ and $\mathbb{R}^3$, we often think of a point (x_1, x_2, x_3, x_4) in $\mathbb{R}^4$ as a vector from the origin, $(0, 0, 0, 0)$, to (x_1, x_2, x_3, x_4). As before, this vector is also denoted (x_1, x_2, x_3, x_4). We can define perpendicularity of vectors in $\mathbb{R}^4$ by extending the characterization that we obtained in the two and three-dimensional cases (Theorems 14.1.2 and 14.2.2).

Definition 14.3.2. The vectors (x_1, x_2, x_3, x_4) and (y_1, y_2, y_3, y_4) are *perpendicular* if $x_1 y_1 + x_2 y_2 + x_3 y_3 + x_4 y_4 = 0$.

Once we have defined four-dimensional space in terms of four-tuples of numbers, it is natural to define five-dimensional space in terms of five-tuples of numbers, and seventeen-dimensional space in terms of seventeen-tuples of numbers. In fact, for every natural number n, n-dimensional Euclidean space can be defined in terms of n-tuples of real numbers. (One-dimensional Euclidean space is simply the real numbers.)

Definition 14.3.3. For each natural number n, *n-dimensional Euclidean space*, denoted $\mathbb{R}^n$, is the set of all n-tuples of real numbers $(x_1, x_2, \ldots, x_n)$ with the distance between the n-tuples $(x_1, x_2, \ldots, x_n)$ and $(y_1, y_2, \ldots, y_n)$ defined to be

$$\sqrt{(x_1 - y_1)^2 + (x_2 - y_2)^2 + \cdots + (x_n - y_n)^2}$$

The elements of n-dimensional space are also called *points* or *vectors*.

Definition 14.3.4. In $\mathbb{R}^n$, the *zero vector* is the vector whose coordinates are all 0. We often use the notation 0 to denote the zero vector; it is apparent from the context whether 0 refers to the number 0 or to the vector 0.

We define perpendicularity for vectors in $\mathbb{R}^n$ by extending the characterization that we obtained in the two and three-dimensional cases (Theorems 14.1.2 and 14.2.2).

Definition 14.3.5. The vectors $(x_1, x_2, \ldots, x_n)$ and $(y_1, y_2, \ldots, y_n)$ are *perpendicular* if $x_1 y_1 + x_2 y_2 + \cdots + x_n y_n = 0$.

We next discuss some properties of n-dimensional spaces.

14.4 Norms and Inner Products

In some spaces, there is a way of capturing the idea of the angle between two vectors. We begin by discussing this concept in $\mathbb{R}^2$. Some preliminary definitions are required.

Definition 14.4.1. The *inner product* (sometimes called the *scalar product* or *dot product*) of the vectors $x = (x_1, x_2)$ and $y = (y_1, y_2)$ in $\mathbb{R}^2$ is $x_1 y_1 + x_2 y_2$. The inner product of the vectors x and y is denoted $\langle x, y \rangle$. That is, $\langle x, y \rangle = x_1 y_1 + x_2 y_2$.

Note that if $x = (x_1, x_2)$, then $\langle x, x \rangle = x_1^2 + x_2^2$, which is the square of the distance from $(0, 0)$ to (x_1, x_2).

Definition 14.4.2. The *norm*, or *length*, of the vector $x = (x_1, x_2)$ in $\mathbb{R}^2$ is $\sqrt{\langle x, x \rangle} = \sqrt{x_1^2 + x_2^2}$. It is denoted $\|x\|$.

Thus, $\|x\|$ is the distance from the origin $(0, 0)$ to the point (x_1, x_2).

Theorem 14.4.3. *For x and y in $\mathbb{R}^2$,*

$$\langle x, y \rangle = \|x\| \cdot \|y\| \cos \theta$$

where θ is the angle between the vectors x and y.

Proof. Since we regard the zero vector as being perpendicular to all vectors, the formula holds if x or y is the zero vector. So assume that x and y are non-zero vectors. (Note that the angle between two non-zero vectors is between 0 and π radians, equivalently, between 0 and 180 degrees.)

Fig. 14.6 Angle between vectors

Consider Figure 14.6. The lengths of the sides of the angle θ are $\|x\|$ and $\|y\|$. Let d be the distance between x and y. The Law of Cosines (Problem 16 in Chapter 11) gives $d^2 = \|x\|^2 + \|y\|^2 - 2\|x\| \cdot \|y\| \cos \theta$. Rearranging the terms in the equation gives

$$\|x\| \cdot \|y\| \cos \theta = \frac{1}{2} \left(\|x\|^2 + \|y\|^2 - d^2 \right)$$

In terms of the coordinates of x and y, the right-hand side of this equation is equal to

$$\frac{1}{2} \left(x_1^2 + x_2^2 + y_1^2 + y_2^2 - \left[(x_1 - y_1)^2 + (x_2 - y_2)^2 \right] \right)$$

We must show that this is equal to $\langle x, y \rangle$. First, it is equal to

$$\frac{1}{2}\left(x_1^2 + x_2^2 + y_1^2 + y_2^2 - \left[x_1^2 - 2x_1y_1 + y_1^2 + x_2^2 - 2x_2y_2 + y_2^2\right]\right)$$

which reduces to

$$\frac{1}{2}(2x_1y_1 + 2x_2y_2) = x_1y_1 + x_2y_2$$

Since $x_1y_1 + x_2y_2 = \langle x, y \rangle$, this gives $\|x\| \cdot \|y\| \cos\theta = \langle x, y \rangle$. $\qquad\square$

Similarly, in $\mathbb{R}^3$ the *norm* of the vector $x = (x_1, x_2, x_3)$ is defined to be $\sqrt{x_1^2 + x_2^2 + x_3^2}$ and is denoted $\|x\|$. The *inner product* of the vectors $x = (x_1, x_2, x_3)$ and $y = (y_1, y_2, y_3)$ is defined to be $x_1y_1 + x_2y_2 + x_3y_3$ and is denoted $\langle x, y \rangle$. As in $\mathbb{R}^2$, it can be proven that in $\mathbb{R}^3$, $\langle x, y \rangle = \|x\| \cdot \|y\| \cos\theta$, where θ is the angle between the vectors x and y (see Problem 9).

Note that in both $\mathbb{R}^2$ and $\mathbb{R}^3$ the fact that the vector x is perpendicular to the vector y if and only if $\langle x, y \rangle = 0$ (Theorems 14.1.2 and 14.2.2) is a special case of the above, since, for θ between 0 and π, $\cos\theta = 0$ if and only if $\theta = \frac{\pi}{2}$ (which is 90°).

Similar definitions can be made in every $\mathbb{R}^n$.

Definition 14.4.4. For vectors $x = (x_1, x_2, \ldots, x_n)$ and $y = (y_1, y_2, \ldots, y_n)$ in $\mathbb{R}^n$, the *inner product* of x and y is $x_1y_1 + x_2y_2 + \cdots + x_ny_n$; it is denoted $\langle x, y \rangle$. The *norm* of the vector x, denoted $\|x\|$, is

$$\sqrt{\langle x, x \rangle} = \sqrt{x_1^2 + x_2^2 + \cdots + x_n^2}$$

There are two basic operations on $\mathbb{R}^n$, addition of vectors and multiplication of vectors by real numbers.

Definition 14.4.5. For vectors $x = (x_1, x_2, \ldots, x_n)$ and $y = (y_1, y_2, \ldots, y_n)$ in $\mathbb{R}^n$, the *sum* of x and y, denoted $x + y$, is the vector $(x_1 + y_1, x_2 + y_2, \ldots, x_n + y_n)$. For t a real number, the *product* of t and the vector x is $(tx_1, tx_2, \ldots, tx_n)$ and is denoted tx. The vector $x - y$ is defined to be $x + (-1)y$, which is, in terms of its coordinates, $(x_1 - y_1, x_2 - y_2, \ldots, x_n - y_n)$.

Note that the distance between the vectors x and y is $\|x - y\|$ (Definition 14.3.3).

There are some important relationships between these operations and norms and inner products.

Theorem 14.4.6 (Properties of Inner Products). *Let x, y, and z be vectors in $\mathbb{R}^n$ and let t be a real number. Then:*

(i) $\langle x, y \rangle = \langle y, x \rangle$
(ii) $\langle tx, y \rangle = t \langle x, y \rangle$
(iii) $\langle x + y, z \rangle = \langle x, z \rangle + \langle y, z \rangle$

Proof. Each of the above is easily verified by simply writing out both sides of the equations in terms of the coordinates of the vectors.   □

It follows immediately from properties (i) and (ii) that $\langle x, ty \rangle = t \langle x, y \rangle$, and from properties (i) and (iii) that $\langle x, y + z \rangle = \langle x, y \rangle + \langle x, z \rangle$.

Theorem 14.4.7 (Properties of Norms). *Let x be a vector in $\mathbb{R}^n$ and let t be a real number. Then:*

(i) $\|x\| \geq 0$
(ii) $\|x\| = 0$ *if and only if x is the zero vector*
(iii) $\|tx\| = |t| \cdot \|x\|$

Proof. The above follow very simply upon writing the norm of the vector x in terms of the coordinates of x.   □

The following inequality is important. It has many known proofs. The proof that we present is probably the simplest to verify (although it may not be the simplest to motivate).

The Cauchy–Schwarz Inequality 14.4.8. *For any vectors x and y in $\mathbb{R}^n$,*

$$|\langle x, y \rangle| \leq \|x\| \cdot \|y\|$$

Proof. This inequality is obviously true if y is the zero vector, since in that case both sides are 0. In the proof that follows we assume that y is not the zero vector (and thus, by Theorem 14.4.7, that $\|y\|^2 \neq 0$).

For every real number t, $\|x - ty\|^2 \geq 0$ (Theorem 14.4.7(i)). As we now show, the theorem follows by applying this inequality with a cleverly chosen t and using the properties of norm and inner product that are listed in Theorems 14.4.6 and 14.4.7.

First, note that

$$\|x - ty\|^2 = \langle x - ty, x - ty \rangle$$
$$= \langle x - ty, x \rangle + \langle x - ty, -ty \rangle$$
$$= \langle x, x \rangle - \langle ty, x \rangle + \langle x, -ty \rangle + \langle -ty, -ty \rangle$$
$$= \|x\|^2 - 2t\langle x, y \rangle + t^2 \|y\|^2$$

We know that this quantity is nonnegative for every real number t. Using this fact for $t = \frac{\langle x, y \rangle}{\|y\|^2}$ gives the inequality

$$0 \leq \|x\|^2 - 2\frac{\langle x, y\rangle}{\|y\|^2}\langle x, y\rangle + \frac{\langle x, y\rangle^2}{\|y\|^4}\|y\|^2 = \|x\|^2 - 2\frac{\langle x, y\rangle^2}{\|y\|^2} + \frac{\langle x, y\rangle^2}{\|y\|^2}$$

Thus,

$$0 \leq \|x\|^2 - \frac{\langle x, y\rangle^2}{\|y\|^2}$$

or

$$\frac{\langle x, y\rangle^2}{\|y\|^2} \leq \|x\|^2$$

Therefore, $\langle x, y\rangle^2 \leq \|x\|^2 \cdot \|y\|^2$, so $|\langle x, y\rangle| \leq \|x\| \cdot \|y\|$.   □

The Cauchy–Schwarz Inequality has a number of important applications. One is in proving a crucial property of the norm.

Theorem 14.4.9 (The Triangle Inequality for Vectors).   *If x and y are vectors in* $\mathbb{R}^n$, *then* $\|x + y\| \leq \|x\| + \|y\|$.

Proof. This follows from the Cauchy–Schwarz Inequality (14.4.8) and the properties of the inner product (Theorem 14.4.6) by an easy direct computation, as follows:

$$\begin{aligned}
\|x + y\|^2 &= \langle x + y, x + y\rangle \\
&= \langle x + y, x\rangle + \langle x + y, y\rangle \\
&= \langle x, x\rangle + \langle y, x\rangle + \langle x, y\rangle + \langle y, y\rangle \\
&= \langle x, x\rangle + 2\langle x, y\rangle + \langle y, y\rangle \\
&\leq \|x\|^2 + 2\|x\| \cdot \|y\| + \|y\|^2 \\
&= (\|x\| + \|y\|)^2
\end{aligned}$$

Therefore, $\|x + y\| \leq \|x\| + \|y\|$.   □

The above is called the "Triangle Inequality" because, in $\mathbb{R}^2$ or $\mathbb{R}^3$, it has the interpretation that the sum of the lengths of two sides of a triangle is greater than or equal to the length of the third side.

14.5   Infinite-Dimensional Spaces

Mathematicians have created a number of different infinite-dimensional spaces. We only discuss a few limited aspects of several such spaces. It seems natural to begin by considering the collection of all sequences of real numbers.

Example 14.5.1. Let S denote the collection of all sequences $(x_1, x_2, x_3, x_4, \ldots)$ where each x_i is a real number.

(Note that the "$\ldots$" means that the sequence continues indefinitely; i.e., there is a term corresponding to each natural number.)

For example, $(\sqrt{2}, \sqrt{3}, \sqrt{4}, \sqrt{5}, \ldots)$ and $(1, -1, 1, -1, \ldots)$ are points in S. While the space of all sequences may have some uses, its utility is limited by the fact that the distance between points cannot be defined in a way analogous to the definition in $\mathbb{R}^n$. We would want the distance between the points $(x_1, x_2, x_3, \ldots)$ and $(y_1, y_2, y_3, \ldots)$ to be the square root of $(x_1 - y_1)^2 + (x_2 - y_2)^2 + (x_3 - y_3)^2 + \cdots$. This would only make sense if $(x_1 - y_1)^2 + (x_2 - y_2)^2 + (x_3 - y_3)^2 + \cdots$ is a convergent series (see Definition 13.1.4). For most sequences $(x_1, x_2, x_3, \ldots)$ and $(y_1, y_2, y_3, \ldots)$, however, this series will not converge. For example, the distance between $(0, 0, 0, \ldots)$ and $(1, 1, 1, \ldots)$ would not be defined.

Though we cannot use this definition of distance for all sequences in S, we can use it for all sequences in some "smaller" infinite-dimensional spaces.

Example 14.5.2. Let $\mathcal{F}$ denote the collection of sequences of real numbers with only a finite number of nonzero terms. That is, $\mathcal{F}$ consists of the set of all sequences $(x_1, x_2, \ldots, x_n, 0, 0, 0, \ldots)$ for natural numbers n and real numbers x_i. Define the distance between $(x_1, x_2, \ldots, x_n, 0, 0, 0, \ldots)$ and $(y_1, y_2, \ldots, y_m, 0, 0, 0, \ldots)$ to be the square root of $(x_1 - y_1)^2 + (x_2 - y_2)^2 + (x_3 - y_3)^2 + \cdots$. Since only a finite number of terms of this sum are different from 0, the distance is finite.

Some of the other basic definitions in $\mathbb{R}^n$ can be extended to the space $\mathcal{F}$. We define the *sum* of two elements of $\mathcal{F}$ to be the coordinate-wise sum. That is, the sum of $(x_1, x_2, \ldots, x_n, 0, 0, 0, \ldots)$ and $(y_1, y_2, \ldots, y_m, 0, 0, 0, \ldots)$ is defined as $(x_1 + y_1, x_2 + y_2, x_3 + y_3, \ldots)$. The *product* of an element of $\mathcal{F}$ by a real number t is defined to be the coordinate-wise product. In other words, the product of t and $(x_1, x_2, \ldots, x_n, 0, 0, 0, \ldots)$ is $(tx_1, tx_2, \ldots, tx_n, 0, 0, 0, \ldots)$.

The *inner product* $\langle x, y \rangle$ of the elements $x = (x_1, x_2, \ldots, x_n, 0, 0, 0, \ldots)$ and $y = (y_1, y_2, \ldots, y_m, 0, 0, 0, \ldots)$ is defined to be $x_1 y_1 + x_2 y_2 + x_3 y_3 + \cdots$. This sum is finite since each of x and y has only a finite number of nonzero coordinates. The *norm* of x is then defined to be $\sqrt{\langle x, x \rangle}$; it is denoted by $\|x\|$.

Infinite-dimensional spaces have many applications in mathematics. The more important infinite-dimensional spaces have a property called "completeness," which we shall not discuss. The space $\mathcal{F}$ is not complete.

An infinite-dimensional space that is much more useful than $\mathcal{F}$ is the following one. It contains more than just the finite sequences but does not contain nearly all sequences. The space we now define is a prototypical example of what is called a *Hilbert space*. This space, and slight variants of it (in particular, using complex numbers rather than real numbers), are very important in mathematics and some areas of physics (including quantum mechanics).

Definition 14.5.3. The space ℓ^2 consists of the set of all sequences of real numbers such that the sum of the squares of the terms of each sequence converges. That is, a sequence $x = (x_1, x_2, x_3, \dots)$ is in ℓ^2 if $x_1^2 + x_2^2 + x_3^2 + \cdots$ converges.

The elements of ℓ^2 are referred to as *vectors* or *points* in ℓ^2. The *zero vector*, denoted by 0, is the sequence in ℓ^2 whose coordinates are all zero. The norm on ℓ^2 is defined as follows.

Definition 14.5.4. For $x = (x_1, x_2, x_3, \dots)$ in ℓ^2, the *norm* of x, denoted $\|x\|$, is
$$\sqrt{x_1^2 + x_2^2 + x_3^2 + \cdots}$$

To establish some basic properties of ℓ^2, we need the following lemma.

Lemma 14.5.5. *If $(x_1, x_2, x_3, \dots)$ and $(y_1, y_2, y_3, \dots)$ are in ℓ^2, then the infinite series $x_1 y_1 + x_2 y_2 + x_3 y_3 + \cdots$ converges.*

Proof. The indicated series converges if it converges absolutely (Problem 15 in Chapter 13). That is, it suffices to show that $|x_1 y_1| + |x_2 y_2| + |x_3 y_3| + \cdots$ converges, and this will follow if it is established that there is an upper bound for the set of partial sums (Theorem 13.5.1).

Let $x = (x_1, x_2, x_3 \dots)$ and $y = (y_1, y_2, y_3 \dots)$. We claim that every partial sum of $|x_1 y_1| + |x_2 y_2| + |x_3 y_3| + \cdots$ is less than or equal to $\|x\| \cdot \|y\|$. To see this, consider any partial sum $S_k = |x_1 y_1| + |x_2 y_2| + \cdots + |x_k y_k|$. The Cauchy–Schwarz Inequality in $\mathbb{R}^k$ (14.4.8) gives $|x_1 y_1| + |x_2 y_2| + \cdots + |x_k y_k| \leq \sqrt{x_1^2 + \cdots + x_k^2} \sqrt{y_1^2 + \cdots + y_k^2}$. Clearly, $\sqrt{x_1^2 + \cdots + x_k^2} \leq \sqrt{x_1^2 + x_2^2 + \cdots}$ and $\sqrt{y_1^2 + \cdots + y_k^2} \leq \sqrt{y_1^2 + y_2^2 + \cdots}$. This implies that the set of partial sums of $|x_1 y_1| + |x_2 y_2| + \cdots$ is bounded by $\|x\| \cdot \|y\|$, and therefore the series converges. $\square$

This lemma allows us to define an inner product on ℓ^2, as follows.

Definition 14.5.6. For vectors $x = (x_1, x_2, x_3, \dots)$ and $y = (y_1, y_2, y_3, \dots)$ in ℓ^2, the *inner product* of x and y, denoted $\langle x, y \rangle$, is defined to be $x_1 y_1 + x_2 y_2 + x_3 y_3 + \cdots$.

As in the finite-dimensional cases we have discussed, there are natural definitions of addition for vectors in ℓ^2 and of multiplication of a vector in ℓ^2 by a real number.

Definition 14.5.7. If $x = (x_1, x_2, x_3, \dots)$ and $y = (y_1, y_2, y_3, \dots)$ are vectors in ℓ^2 and t is a real number, then:

(i) $x + y = (x_1 + y_1, x_2 + y_2, x_3 + y_3, \dots)$
(ii) $tx = (tx_1, tx_2, tx_3, \dots)$

To show that these operations are well-defined on ℓ^2, we must prove that the vectors $x + y$ and tx are in ℓ^2 whenever x and y are in ℓ^2 and t is a real number.

Lemma 14.5.8. *If $x = (x_1, x_2, x_3, \dots)$ and $y = (y_1, y_2, y_3, \dots)$ are vectors in ℓ^2 and t is a real number, then $x + y$ and tx are in ℓ^2.*

Proof. To prove that $x + y$ is in ℓ^2, it must be shown that $(x_1 + y_1)^2 + (x_2 + y_2)^2 + (x_3 + y_3)^2 + \cdots$ converges. For each k, $(x_k + y_k)^2 = x_k^2 + 2x_k y_k + y_k^2$. Since x and y are in ℓ^2, the series $x_1^2 + x_2^2 + x_3^2 + \cdots$ and $y_1^2 + y_2^2 + y_3^2 + \cdots$ both converge. The series $x_1 y_1 + x_2 y_2 + x_3 y_3 + \cdots$ converges by Lemma 14.5.5. Therefore, the series $\left(x_1^2 + 2x_1 y_1 + y_1^2\right) + \left(x_2^2 + 2x_2 y_2 + y_2^2\right) + \cdots$ converges (Theorems 13.3.1 and 13.3.3), so $x + y$ is in ℓ^2.

The series $t^2 x_1^2 + t^2 x_2^2 + t^2 x_3^2 + \cdots$ converges since $x_1^2 + x_2^2 + x_3^2 + \cdots$ does (by Theorem 13.3.1), so tx is in ℓ^2.   □

Theorem 14.5.9 (Properties of the Norm in ℓ^2). *Let x be in ℓ^2 and let t be a real number. Then:*

(i) $\|x\| \geq 0$
(ii) $\|x\| = 0$ *if and only if x is the zero vector*
(iii) $\|tx\| = |t| \cdot \|x\|$

Proof. Each part of this theorem follows easily by writing the norm of the vector x in terms of the coordinates of x. Part (iii) also requires the fact that a term-by-term product of an infinite series by a real number converges to the product of the number and the sum of the original series (Theorem 13.3.1).   □

Theorem 14.5.10 (Properties of the Inner Product in ℓ^2). *The inner product on ℓ^2 satisfies the following properties:*

(i) $\langle x, y \rangle = \langle y, x \rangle$
(ii) $\langle tx, y \rangle = t \langle x, y \rangle$
(iii) $\langle x + y, z \rangle = \langle x, z \rangle + \langle y, z \rangle$
(iv) $\|x\| = \sqrt{\langle x, x \rangle}$

Proof. This theorem follows easily by writing the vectors x and y in terms of their coordinates and using the fundamental properties of infinite series (see Theorems 13.3.1 and 13.3.3).   □

Theorem 14.5.11 (The Cauchy–Schwarz Inequality in ℓ^2). *If x and y are in ℓ^2, then $|\langle x, y \rangle| \leq \|x\| \cdot \|y\|$.*

Proof. The proof of the Cauchy–Schwarz Inequality given for $\mathbb{R}^n$ (Theorem 14.4.8) depends only on the properties of inner product and norm listed in Theorems 14.4.6 and 14.4.7. Since these properties also hold for ℓ^2 (Theorems 14.5.9 and 14.5.10), the theorem follows.   □

Theorem 14.5.12 (The Triangle Inequality in ℓ^2). *If x and y are in ℓ^2, then*

$$\|x + y\| \leq \|x\| + \|y\|$$

Proof. This follows directly from the Cauchy–Schwarz Inequality and the properties of inner products (Theorem 14.5.10), exactly as in the finite-dimensional case (Theorem 14.4.9).   □

Definition 14.5.13. The *distance between the vectors* $x = (x_1, x_2, x_3, \dots)$ and $y = (y_1, y_2, y_3, \dots)$ in ℓ^2 is $\|x - y\|$. That is, the distance between the vectors x and y in ℓ^2 is

$$\sqrt{(x_1 - y_1)^2 + (x_2 - y_2)^2 + (x_3 - y_3)^2 + \cdots}$$

The definition of perpendicularity in ℓ^2 is analogous to the definition in finite-dimensional spaces.

Definition 14.5.14. The vectors x and y in ℓ^2 are *perpendicular* (or *orthogonal*) if $\langle x, y \rangle = 0$.

Some of the basic geometry of the plane and three-dimensional space has analogues in ℓ^2.

Theorem 14.5.15 (The Pythagorean Theorem in ℓ^2). *If x and y are in ℓ^2 and x is perpendicular to y, then $\|x + y\|^2 = \|x\|^2 + \|y\|^2$.*

Proof. Using the basic properties of the inner product on ℓ^2 (Theorem 14.5.10) gives

$$\|x + y\|^2 = \langle x + y, x + y \rangle$$
$$= \langle x, x \rangle + \langle x, y \rangle + \langle y, x \rangle + \langle y, y \rangle$$
$$= \|x\|^2 + 2\langle x, y \rangle + \|y\|^2$$

Since x and y are perpendicular, $\langle x, y \rangle = 0$ (Definition 14.5.14), and so the result follows. □

The following theorem can be interpreted as stating that the sum of the squares of the lengths of the two diagonals of a parallelogram is equal to the sum of the squares of the lengths of its four sides.

Theorem 14.5.16 (The Parallelogram Law). *For vectors x and y in ℓ^2,*

$$\|x + y\|^2 + \|x - y\|^2 = 2\left(\|x\|^2 + \|y\|^2\right)$$

Proof. The relationship between the inner product and the norm (Theorem 14.5.10(iv)) gives

$$\|x + y\|^2 + \|x - y\|^2 = \langle x + y, x + y \rangle + \langle x - y, x - y \rangle$$

Expanding the right-hand side yields

$$\langle x, x \rangle + 2\langle x, y \rangle + \langle y, y \rangle + \langle x, x \rangle - 2\langle x, y \rangle + \langle y, y \rangle$$

which equals $2\langle x, x \rangle + 2\langle y, y \rangle = 2\|x\|^2 + 2\|y\|^2$, as desired. □

14.6 A Difference Between Finite and Infinite-Dimensional Spaces

As we have seen, some of the properties of the infinite-dimensional space ℓ^2 are entirely analogous to the corresponding properties of finite-dimensional Euclidean spaces. On the other hand, there are a number of important differences. We will discuss one of them.

Definition 14.6.1. In ℓ^2, or in any finite-dimensional Euclidean space, if y is a vector and r is a positive number, the *closed ball with center y and radius r* is the set of all vectors, x, whose distance from y is less than or equal to r; that is, $\{x : \|x - y\| \le r\}$.

In $\mathbb{R}^2$, a closed ball is a circle together with its interior (see Figure 14.7). In $\mathbb{R}^3$, a closed ball is a sphere together with its interior.

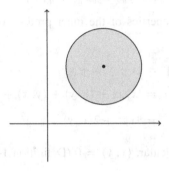

Fig. 14.7 A closed ball in $\mathbb{R}^2$

How many balls of a given radius r can be inside a ball of radius R without intersecting each other? Using the areas of the balls in $\mathbb{R}^2$, or the volumes of the balls in $\mathbb{R}^3$, would give certain limits; the total of the areas (or volumes) of the smaller balls can't exceed the area (or volume) of the larger ball. There are similar limitations in $\mathbb{R}^n$ for any natural number n.

Theorem 14.6.2. *In any $\mathbb{R}^n$, given a closed ball B and an $r > 0$, there cannot be an infinite collection of non-intersecting balls of radius r that are all contained in B.*

Proof. Let $R > 0$ be the radius of the ball B. We start by assuming that B is centered at the origin. As we show below, the general case follows easily from this special case.

To prove the case where B is centered at the origin, we first choose a finite set, S, of points in $\mathbb{R}^n$ such that every point in B is within r of at least one point in S. We then show that if there are more balls of radius r in B than points in S, then two of the balls must intersect.

For this purpose, choose a natural number m such that $m > R \cdot \frac{\sqrt{n}}{r}$. Let the set S be defined as

$$S = \left\{ \left(\frac{k_1 r}{\sqrt{n}}, \dots, \frac{k_n r}{\sqrt{n}} \right) : \text{each } k_i \text{ is an integer with } |k_i| \leq m \right\}$$

Note that S is a finite set, since there are only finitely many integers with absolute value at most m.

If t is any real number, there is some integer a such that $a \leq \frac{t\sqrt{n}}{r} < a + 1$. Thus, $\frac{ar}{\sqrt{n}} \leq t < \frac{(a+1)r}{\sqrt{n}}$ from which it follows that $\left| t - \frac{ar}{\sqrt{n}} \right| < \frac{r}{\sqrt{n}}$. Therefore, if $x = (x_1, \dots, x_n)$ is any point in B, then for each x_i there is an integer a_i such that $\left| x_i - \frac{a_i r}{\sqrt{n}} \right| < \frac{r}{\sqrt{n}}$. This inequality implies that

$$\left| \frac{a_i r}{\sqrt{n}} \right| = \left| \frac{a_i r}{\sqrt{n}} - x_i + x_i \right| \leq \left| \frac{a_i r}{\sqrt{n}} - x_i \right| + |x_i| < \frac{r}{\sqrt{n}} + |x_i|$$

Since no x_i can have absolute value greater than R (otherwise $\|x\|$ would be greater than R) and $\frac{mr}{\sqrt{n}}$ is greater than R,

$$\left| \frac{a_i r}{\sqrt{n}} \right| < \frac{r}{\sqrt{n}} + R < \frac{r}{\sqrt{n}} + \frac{mr}{\sqrt{n}} = (m+1)\frac{r}{\sqrt{n}}.$$

Dividing through by $\frac{r}{\sqrt{n}}$ implies that $|a_i| < m + 1$; that is, $|a_i| \leq m$. It follows that the point $p = \left(\frac{a_1 r}{\sqrt{n}}, \dots, \frac{a_n r}{\sqrt{n}} \right)$ is in S and

$$\|x - p\| = \left\| (x_1, \dots, x_n) - \left(\frac{a_1 r}{\sqrt{n}}, \dots, \frac{a_n r}{\sqrt{n}} \right) \right\|$$

$$= \sqrt{ \left(x_1 - \frac{a_1 r}{\sqrt{n}} \right)^2 + \cdots + \left(x_n - \frac{a_n r}{\sqrt{n}} \right)^2 }$$

$$< \sqrt{ n \cdot \left(\frac{r}{\sqrt{n}} \right)^2 } = \sqrt{r^2} = r$$

This shows that every ball of radius r in B contains a point in S. Thus, if $\mathcal{T}$ is a collection of balls of radius r, each of which is contained in B, and has more balls than points in S, then two of the balls in $\mathcal{T}$ must contain the same point from S, and so they intersect. Hence, there cannot be infinitely many non-intersecting balls of radius r within a ball of radius R centered at 0.

To prove the case where the given ball is not centered at the origin, suppose there are infinitely many non-intersecting balls of radius r contained in a ball B of radius R centered at a point y. We establish this general case by "translating" to the case where the ball is centered at the origin. That is, subtracting y from all the points in each of the balls would give infinitely many non-intersecting balls of radius r contained in a ball of radius R centered at 0. This would contradict the previous case. $\square$

In ℓ^2, however, an infinite number of non-intersecting closed balls of a fixed radius can be contained in a given closed ball.

Example 14.6.3. In ℓ^2, every closed ball of radius 2 contains an infinite collection of closed balls of radius $\frac{1}{3}$, no two of which intersect.

Proof. We begin by considering the particular case of the ball of radius 2 centered at the origin, which we denote by B. That is, B is the set of all vectors in ℓ^2 whose norms are less than or equal to 2. For each natural number k, let e_k be the vector in ℓ^2 whose k^{th} coordinate is 1 and all of whose other coordinates are 0. Note that, for every k, the norm of e_k is 1. Define the ball B_k to be the set of all vectors that are at most $\frac{1}{3}$ from e_k. That is, $B_k = \left\{ x \in \ell^2 : \|x - e_k\| \le \frac{1}{3} \right\}$. Thus, $\left\{ B_k : k \in \mathbb{N} \right\}$ is an infinite set of balls of radius $\frac{1}{3}$. We now show that each B_k is contained in B and that no two distinct B_k's intersect.

If x is in B_k, then, by the Triangle Inequality (14.5.12), $\|x\| = \|x - e_k + e_k\| \le \|x - e_k\| + \|e_k\| \le \frac{1}{3} + 1 < 2$. Thus, B_k is contained in B for every k.

To show that no two B_k's intersect, suppose that x was in both B_i and B_j, where i and j are distinct. Then $\|x - e_i\| \le \frac{1}{3}$ and $\|x - e_j\| \le \frac{1}{3}$, from which it follows, using the Triangle Inequality (14.5.12), that

$$\|e_i - e_j\| = \|e_i - x + x - e_j\| \le \|e_i - x\| + \|x - e_j\| \le \frac{1}{3} + \frac{1}{3} < 1$$

This contradicts the fact that $\|e_i - e_j\| = \sqrt{1 + 1} = \sqrt{2}$, which is greater than 1. Therefore no x can be in two distinct B_k's, so no two B_k's intersect.

To establish the case where the ball is not centered at the origin, let C be a closed ball of radius 2 centered at the vector x_0. For each natural number k, let C_k equal the set of all vectors $x + x_0$ such that x is in B_k. It is straightforward to verify that every C_k is a closed ball of radius $\frac{1}{3}$ that is contained in C and that the C_k are non-intersecting. $\square$

14.7   Problems

Basic Exercises

1. Which of the following pairs of vectors in $\mathbb{R}^2$ are perpendicular to each other?

   (a) $(0, 1)$ and $(1, 0)$
   (b) $(7, \pi)$ and $(\pi, -7)$
   (c) $\left(\sqrt{3}, \sqrt{22} \right)$ and $\left(\sqrt{22}, -3 \right)$

2. Which of the following pairs of vectors in $\mathbb{R}^3$ are perpendicular to each other?

   (a) $\left(\frac{1}{2}, \frac{1}{2}, 2\right)$ and $\left(4, 6, -\frac{5}{7}\right)$
   (b) $\left(-4, \sqrt{2}, 58\right)$ and $\left(58, 4, \sqrt{2}\right)$

3. Suppose that x and y are vectors in $\mathbb{R}^2$ whose coordinates are all positive. Show that x is not perpendicular to y.

4. Suppose that the vectors x and y in $\mathbb{R}^n$ are both perpendicular to the vector z. Show that $x + y$ is perpendicular to z.

5. In ℓ^2, for each natural number k, let e_k be the vector whose k^{th} coordinate is 1 and whose other coordinates are all 0. Prove that e_i is perpendicular to e_j whenever i is different from j.

Interesting Problems

6. Suppose that x and y are vectors in $\mathbb{R}^3$ whose coordinates are all positive. Show that x is not perpendicular to y.

7. Show that, in $\mathbb{R}^n$ and in ℓ^2, the vectors $x + y$ and $x - y$ are perpendicular to each other if and only if $\|x\| = \|y\|$.

8. The space ℓ^1 is defined to be the set of all sequences of real numbers $(x_1, x_2, x_3, \ldots)$ such that $|x_1| + |x_2| + |x_3| + \cdots$ converges. The *norm* of the vector $x = (x_1, x_2, x_3, \ldots)$ in ℓ^1 is defined to be $|x_1| + |x_2| + |x_3| + \cdots$ and is denoted $\|x\|$. The sum of the vectors $(x_1, x_2, x_3, \ldots)$ and $(y_1, y_2, y_3, \ldots)$ is defined to be $(x_1 + y_1, x_2 + y_2, x_3 + y_3, \ldots)$. The product of the vector $(x_1, x_2, x_3, \ldots)$ by the real number t is $(tx_1, tx_2, tx_3, \ldots)$. Suppose that x and y are in ℓ^1 and t is a real number. Prove that $x + y$ is in ℓ^1, tx is in ℓ^1, and $\|x + y\| \leq \|x\| + \|y\|$.

9. Prove that, for all vectors x and y in $\mathbb{R}^3$, $\langle x, y \rangle = \|x\| \cdot \|y\| \cos \theta$, where θ is the angle between the vectors x and y.
   [Hint: Use the Law of Cosines as in the proof of Theorem 14.4.3.]

Challenging Problems

10. Define *n-dimensional complex space*, denoted $\mathbb{C}^n$, to be the set of all n-tuples of complex numbers, where the *norm* of the n-tuple $(x_1, x_2, \ldots, x_n)$ is $\sqrt{|x_1|^2 + |x_2|^2 + \cdots + |x_n|^2}$. The sum of two vectors is defined coordinate-wise, as is multiplication of a vector by a complex number. Define the *inner product* of the vectors $x = (x_1, x_2, \ldots, x_n)$ and $y = (y_1, y_2, \ldots, y_n)$ to be $\langle x, y \rangle = x_1 \overline{y_1} + x_2 \overline{y_2} + \cdots + x_n \overline{y_n}$. (Recall that, for any complex number $z = a + bi$, the complex conjugate $\overline{z}$ is $a - bi$.)

    (a) Prove the Cauchy–Schwarz Inequality: $|\langle x, y \rangle| \leq \|x\| \cdot \|y\|$.
    (b) Prove the Triangle Inequality: $\|x + y\| \leq \|x\| + \|y\|$

11. The norm on ℓ^1 (defined in Problem 8) would be said to "arise from an inner product" if there is a function taking pairs of vectors in ℓ^1 to the real numbers that satisfies the properties of the inner product listed in Theorem 14.5.10. Prove that the norm on ℓ^1 does not arise from an inner product.

[Hint: One way to do this is the following. First prove that if the norm did arise from an inner product, then the norm would satisfy the "Parallelogram Law" as stated in Theorem 14.5.16. Then find a pair of vectors in ℓ^1 for which the Parallelogram Law fails.]

12. Generalize Example 14.6.3 to prove:

(a) If $R > 0$ and B is a closed ball in ℓ^2 of radius R, then there is an $r > 0$ such that B contains an infinite collection of non-intersecting closed balls of radius r.

(b) If $R > 0$ and B is a closed ball in ℓ^1 of radius R, then there is an $r > 0$ such that B contains an infinite collection of non-intersecting closed balls of radius r.

13. (This problem requires the basics of integral calculus.) For f and g continuous functions from $[0, 1]$ to $\mathbb{R}$, define $\langle f, g \rangle = \int_0^1 f(t)g(t)dt$ and define $\|f\| = \left(\int_0^1 |f(t)|^2 dt \right)^{1/2}$

(a) Prove that $\langle f, g \rangle$ has the properties of an inner product listed in Theorem 14.5.10.

(b) Prove that $\|f\|$ has the properties of a norm listed in Theorem 14.5.9.

(c) Prove that

$$\left| \int_0^1 f(t)g(t)dt \right| \leq \left(\int_0^1 |f(t)|^2 dt \right)^{1/2} \left(\int_0^1 |g(t)|^2 dt \right)^{1/2}$$

(d) Prove that

$$\left(\int_0^1 |f(t) + g(t)|^2 dt \right)^{1/2} \leq \left(\int_0^1 |f(t)|^2 dt \right)^{1/2} + \left(\int_0^1 |g(t)|^2 dt \right)^{1/2}$$

14. Let $\mathcal{V}$ denote any one of the spaces $\mathbb{R}^n$, ℓ^1, or ℓ^2. Let T be a *linear transformation* from $\mathcal{V}$ to itself. That is, T is a function from $\mathcal{V}$ to $\mathcal{V}$ such that: (i) $T(x+y) = T(x)+T(y)$ for all vectors x and y in $\mathcal{V}$; and (ii) $T(rx) = rT(x)$ for all real numbers r and all vectors x in $\mathcal{V}$.

(a) Prove that $T(0) = 0$.

(b) Prove that T is one-to-one (see Definition 10.1.5) if and only if the only vector that satisfies $T(x) = 0$ is the zero vector.

15. Let $\mathcal{V}$ denote any one of the spaces $\mathbb{R}^n$, ℓ^1, or ℓ^2. As in calculus, a function F taking $\mathcal{V}$ into $\mathcal{V}$ is said to be *continuous at the vector a* if, for every $\varepsilon > 0$ there is a $\delta > 0$ such that $\|F(x) - F(a)\| < \varepsilon$ whenever $\|x - a\| < \delta$.

   (a) Prove that a linear transformation T is continuous at every vector a in $\mathcal{V}$ if and only if T is continuous at 0.
   (b) A linear transformation T is said to be *bounded* if there exists a positive number K such that $\|T(x)\| \leq K\|x\|$ for every vector x. Prove that T is continuous at every a if and only if T is bounded.
   (c) Prove that, for every n, every linear transformation from $\mathbb{R}^n$ to itself is continuous at every vector in $\mathbb{R}^n$.

Index

Printed in the United States
by Baker & Taylor Publisher Services